Principles and Practice of Forensic Mechanical Engineering

Principles and Practice of Forensic Mechanical Engineering explains the physical processes and causes of failures that result in physical or bodily damage. It discusses the nature of defects and defines the three types – design defects, manufacturing defects and service defects – with supporting examples of each.

Drawing on the author's decades of experience, the book advises mechanical engineers how to investigate the cause of an incident, evaluate responsibility and prepare a report of the results. Rather than focusing solely on theoretical aspects of physics or chemistry, readers will learn how physical and chemical principles can be applied to solve real-world cases. Through numerous examples, the book emphasizes the need for not only technical knowledge but also thinking outside the box to resolve challenging cases.

This book supplies essential knowledge about forensic procedures to help mechanical engineers become involved in and efficiently practice forensics.

Principles and Practice of Forensic Mechanical Engineering

Fredric M. Blum

CRC Press
Taylor & Francis Group
Boca Raton London New York

CRC Press is an imprint of the
Taylor & Francis Group, an **informa** business

First edition published 2026
by CRC Press
2385 NW Executive Center Drive, Suite 320, Boca Raton FL 33431

and by CRC Press
4 Park Square, Milton Park, Abingdon, Oxon, OX14 4RN

CRC Press is an imprint of Taylor & Francis Group, LLC

ISBN: 978-1-041-16367-1 (hbk)
ISBN: 978-1-041-16368-8 (pbk)
ISBN: 978-1-003-68421-3 (ebk)

DOI: 10.1201/9781003684213

Typeset in Times
by Apex CoVantage, LLC

Contents

About the Author

Fredric M. Blum has practiced as a forensic mechanical engineer for more than 46 years. He has investigated thousands of cases on behalf of plaintiffs and defendants, insurance companies and attorneys, written thousands of reports, given deposition testimony numerous times and testified more than 200 times in court. He has lectured at professional meetings and written treatises on various aspects of forensic mechanical engineering.

Mr. Blum received a bachelor's degree in Mechanical Engineering from Drexel University, Philadelphia, PA, in 1963 and a master's degree in Business Administration from the Wharton School of the University of Pennsylvania, Philadelphia, PA, in 1965. He is also a Certified Fire and Explosion Investigator. Mr. Blum was employed for four years as a systems analyst at Auerbach Corp., Philadelphia, PA, and for five years as a systems analyst and Technical Manager at Analytics, Inc., Willow Grove, PA. For four years thereafter, he worked in construction carpentry and founded and operated his own residential design and construction company in Santa Fe, NM. For most of his career as a forensic mechanical engineer, he was affiliated with Paul Zamrowski Associates, West Chester, PA. He is presently an independent practitioner.

Mr. Blum has authored several articles on forensic mechanical engineering, including "Oil Burner Fires" (1980), "Kerosene Heater Fires" (1981) and "Soot Contamination from Oil Burners, Gas Appliances & Candles" (2002).

Acknowledgments

Profound thanks to Paul Zamrowski, Sr, for giving me the opportunity to pursue and blossom in forensic mechanical engineering. When we met, I had never even heard of the field, but he recognized that I was qualified because of my diverse background. Thanks also to the many attorneys, particularly those at Cozen O'Connor, LLP, Philadelphia, PA, many of whom were my clients over the years. They educated me about the law and legal procedures, which are essential for a forensic mechanical engineer in order to provide the most cogent advice to clients based on findings, as well as how findings should best be presented.

I also owe great thanks to my wife, Stephanie Kitain, whose unfailing support made it possible for me to write this book.

Foreword

Many books have been written about forensic engineering. Most deal with structural failures and thus fall under the aegis of forensic civil engineering. Hardly any touch on forensic mechanical engineering, which is one reason I wrote this book. The other reason is to share my 46 years of experience as a forensic mechanical engineer, highlighting the basic principles a successful practitioner must master and interesting examples of actual cases. This is not a textbook, but rather a handbook with examples. Readers seeking a textbook should turn to an introductory textbook of physics or physical chemistry because these are the fields of knowledge upon which forensic mechanical engineering is based. Diversity of knowledge is the prime qualification since it is often impossible to anticipate what sort of knowledge will be necessary to solve a given case.

Introduction 1

Things happen.

Good things, bad things. Bad things like accidents, failures, fires and other catastrophes. They are called "accidents" if someone is injured or "incidents" if there is damage but no injuries.

As a rule, if something *can* happen, sooner or later it *will* happen.

If something happens to others, we may be sympathetic but not care why it happened. Only the people involved may care why. Police and firefighters are paid to care why, but often lack the technical skills needed to find out. Because incidents and accidents often have financial consequences, insurance companies and attorneys need to know why, but they also usually lack the necessary skills. Finding why is the province of forensic engineering.

"Forensic" means having to do with the law. Mechanical engineering is engineering applied to mechanical devices – design, manufacture, maintenance, repair, sales and use. Forensic mechanical engineering is concerned with calamities involving mechanical devices.[1] Calamities may have legal consequences, so a forensic mechanical engineer may have to testify about their findings.

Mechanical equipment heats things, pressurizes things, propels things, transports things, distributes fluids, stores things, cuts things, forms things, etc. Mechanical equipment can fail in myriad ways. Mechanical devices seize up, they leak, they collapse, they freeze, they catch fire and components break. Sometimes, a mechanical device does not fail but is involved in an accident in which someone is injured. Forensic mechanical investigation of so many kinds of incidents involving so many kinds of mechanical devices requires diverse knowledge, including the physics of motion, how machines work, how materials behave, corrosion, freezing, vaporization, fire science, pressure versus temperature, etc. This book describes some of the specialized knowledge a forensic mechanical engineer needs.[2]

DOI: 10.1201/9781003684213-1

REPRESENTATION, ADVOCACY AND IMPARTIALITY

Lawyers have clients. Forensic engineers have clients. Clients pay these specialists for their services. However, the duties and obligations differ between lawyers and forensic engineers. Like other forensic specialties such as pathology, toxicology, fingerprint analysis, genetic analysis, and tire mark analysis, forensic mechanical engineers are independent consultants typically hired on an hourly basis by insurance companies and attorneys. Although a consultant naturally wishes to please the client, it is essential that a forensic investigation be conducted impartially, avoiding bias and holding truth as the guiding star. Otherwise, the engineer's findings could be considered biased and vulnerable to questions of credibility.

A defense attorney is hired to advocate for their client's interest regardless of the client's guilt, innocence, or apparent culpability. A plaintiff's attorney is hired to advocate for their client's case regardless of the merits of the case. In this sense, advocacy by an attorney is analogous to advertising and public relations, which do not necessarily require truthfulness or accuracy. By contrast, a forensic engineer is retained to find out what happened, why it happened and who, if anyone, is responsible. A forensic engineer is an advocate, but only for his results, not for the client's case, regardless of its merits. His results may or may not favor the client's interest. In other words, a forensic engineer's allegiance is only to the truth. A client who wants a forensic engineer to espouse the client's position regardless of its merits misunderstands a forensic engineer's role, where non-partisanship and independence are the foundation of the code of ethics. A forensic engineer has no choice but to decline a case when the prospective client makes it clear in advance that the engineer's findings are expected to support the client's interest.

Notes

1 Determining what happened and why it happened is sometimes called "accident reconstruction." However, for some reason the term is customarily applied only to vehicular accidents.

2 Although automotive and aeronautical engineering are to a large degree mechanical, each is a major specialty of its own and therefore not covered in this book.

Defects 2

THE CAUSE OF A CALAMITY

In forensic engineering parlance, a defect is a fault or imperfection that results in significant adverse economic, medical or legal consequences. Significant defects may cause failures, fires or accidents. Lesser faults or imperfections are just annoyances. Defects may pertain to design, manufacture, sale, service, operating instructions, warnings or user training. The task of a forensic engineer is to identify all significant defects – if any – relevant to an incident, explain how each contributed to the incident, and identify who if anyone is responsible for the defect.

Defects fall into four general categories: design, manufacturing, installation and service.

DESIGN DEFECTS

Product design encompasses a device's form and function as well as instructional material and warnings. A product's form can be efficient and safe or inefficient and unsafe. A product can be durable or unreliable. A product failure caused by poor design can lead to an accident or cause property damage.

We have all experienced poorly designed devices in our daily lives – products that don't work as advertised, products that fail prematurely, products that are difficult to use or difficult to repair. Most bad designs do not lead to catastrophes. Forensic mechanical engineering deals with those that do.

For instance, defective design might be manifest as undersized bolts, too few bolts, lack of a guard, the structure being undersized for expected stresses, violation of a design or safety principle, failure to warn of a hazard, failure to provide necessary operating instructions, etc. The list of types of design defects is almost endless.

DOI: 10.1201/9781003684213-2

The worst catastrophic failures from bad design are reported in the media. Examples include the Titanic whose design incorporated inferior quality hull bolts and too few lifeboats, two 737 Max aircraft whose defectively designed attitude control system caused them to nosedive, the Ocean Ranger drilling rig that sank off the coast of Newfoundland when water entered the control room through a glass porthole that broke in a storm because its steel cover had been left open and lacked a sensor to monitor its status.

Failure to Warn

Instructions and warnings are considered part of the design of a machine or device because proper usage and knowledge of hazards are often crucial for safety. Failure to provide proper operating instructions and hazard warnings which may cause or contribute to an accident or catastrophe therefore constitutes a design defect.

The torch explosion case described in detail elsewhere herein is a good example. In essence, the user was burned when a fireball erupted at a butane-fueled kitchen torch shortly after it was refilled. The instructions provided with the torch did not mention that butane vapor is emitted every time the torch is refilled, or that butane vapor is heavier than air and will sink and potentially flow to a source of ignition, which is what happened in this case. Flame followed the vapor trail back to the torch, where a vapor cloud burst into a fireball. (The fireball appeared to have been emitted from the torch, but it actually had not.) The user had read the instructions which failed to provide any insight about the true hazard of butane vapor. Proper instructions and warnings could have averted the accident by enabling the user to take necessary precautions.

Gas Cooktop Outflash

A homeowner received a new gas cooktop as a gift. The cooktop was a major brand that is marketed directly and as a house brand throughout the country. The cooktop featured four conventional burners, each ignited by a spark when the burner control knob is rotated to the ignite position. The homeowner found that each burner produced a brief, taller-than-normal flame (outflash) when it was turned on. The homeowner thought this seemed abnormal and dangerous because her previous cooktop burners did not produce high-turn-on flames. Therefore, she called the manufacturer for service under the warranty.

A service technician concluded that the set of wires (wiring harness) from the switch on each knob shaft to the spark generator was defective, so he ordered a replacement. A week later, the technician returned and installed the new harness. However, the outflash problem continued.

The homeowner decided to live with the problem, but remained wary, so assuming it would be safer, she turned the knobs more slowly than usual. One day, she placed a pot on the cooktop, holding the handle with her right hand while turning the burner knob with her left hand. As usual, an outflash of flame erupted from when the burner ignited, but this time the flame caught the sleeve of the robe she was wearing. A friend who was present quickly rolled her on the floor to extinguish the fire, but she incurred burns on her arm, neck and face.

Testing by a forensic mechanical engineer retained on behalf of the homeowner confirmed the propensity of all burners to produce an outflash every time a burner was turned on. The original wiring harness, which the homeowner had kept, was also tested and found to function normally. In fact, the outflash phenomenon had nothing to do with the wiring harness. The outflashes happened because when a burner knob was rotated, gas began flowing before the ignition position was reached. Therefore, unburned gas accumulated at the burner, which ignited all at once when a spark appeared. An outflash is a mini-explosion. The homeowner's more cautious (slower) knob rotation exacerbated the problem because it allowed more gas to flow before ignition. The gas-before-ignition control logic violates the most basic safety principle for any sort of combustion appliance which mandates that a source of ignition always be present before fuel is introduced. This rule assures that fuel cannot accumulate before ignition. Ignition-before-gas guarantees that the gas will ignite immediately upon introduction. Ignition of an accumulation of fuel is what causes an explosion. Thus, the engineer concluded that the cooktop was defective by design.

His opinion was valid regardless of how many thousands of this brand of gas cooktops and ranges are sold throughout the country. All are dangerous. Defective is defective, regardless of quantity.

The forensic engineer also checked to see if the cooktop's faulty gas-before-ignition control logic might be standard in the industry, but it is not. Almost all major brands of gas cooking appliances incorporate the opposite, safe, ignition-before-gas control logic.

The cooktop manufacturer was held liable for the accident.

Smelting Furnace Explosion

A large electric induction furnace was used to melt scrap aluminum into aluminum ingots. (An induction furnace uses alternating electric current in a coil surrounding a crucible to induce current in, and thereby heat, heat electrically conductive material in the crucible.) The crucible in which aluminum was melted was situated inside a protective chamber. The crucible was made of steel with a refractory (ceramic) lining. After a batch had melted, the crucible was tilted to pour the aluminum into molds positioned inside the chamber. Crucible tilting was hydraulically powered, with pressurized hydraulic fluid (oil) carried to the hydraulic cylinder by a steel-reinforced hose (not a steel pipe) on the chamber floor. One day during a melt, the operator, looking through a window in the chamber roof, observed that the level of molten aluminum in the crucible had fallen. The crucible was still upright. Realizing that the crucible must have a leak, he shut off power. After allowing the molten aluminum to cool somewhat, the chamber was opened and instantly exploded. The operator suffered burns and the chamber was damaged.

Forensic investigation determined that the explosion was fueled by oil vapor generated when the molten aluminum leaking from the crucible damaged the hydraulic oil line on the chamber floor. The high temperature in the chamber vaporized the oil. Aluminum melts at 1,220 °F and hydraulic fluid ignites at approximately 660–825 °F, so the oil vapor immediately ignited. However, the oxygen supply in the chamber was quickly consumed, thereby leaving a great deal of vapor in the chamber. When the chamber was opened and fresh air (oxygen) was admitted, the vapor exploded because the molten aluminum in the crucible or on the chamber floor was still above the vapor's ignition temperature.

Afterwards, the leak was traced to a wear-out-type crack in the crucible's refractory liner and a loose joint in the steel body. The explosion was deemed to have been caused by the premature opening of the chamber for which the plant was responsible (the procedure was improper or training was inadequate), but the furnace manufacturer was responsible for the defective design of the chamber whereby the vulnerable hydraulic line was run on the floor. The possibility of a crucible leak that would compromise a line on the floor was foreseeable, which should have mandated that the line be made of steel and/or elevated.

Mixer Accident

A large horizontal, cylindrical mixing machine was used to prepare dough in a commercial bakery. Inside the mixer, blades rotated slowly between fixed vanes to mix ingredients. Operators on an adjacent catwalk pour ingredients in via a hatch on the end of the cylinder and the finished product is emptied through the hatch. The mixer was operated from a pushbutton control panel located on the catwalk railing. Periodically, the mixer was cleaned and sanitized by a worker who climbed into the mixer via the hatch and wiped all surfaces. One day, a worker was cleaning inside the mixer when a coworker outside leaned over the catwalk railing to speak with the worker inside. In leaning over the catwalk, the coworker inadvertently depressed the mixer's start button on the control panel. The worker inside was fatally injured by the rotating blades.

Forensic investigation determined that the accident was caused by two design defects. First, the control panel and specifically the start button were unprotected from accidental contact. Second, power to the machine was not supplied through a key-operated master power switch. A key-operated switch would enable a worker to turn off power and keep the key with him so the machine could not be accidentally started while the worker was inside. The machine manufacturer was held responsible for the accident by providing an unsafe control panel and failing to provide instructions for the employer to provide a key-operated power switch.

Deep Fat Turkey Fryers

Deep fat turkey fryers are large, cylindrical pots that sit on propane-fired burners mounted on legs. The burners typically are manually controlled. The user monitors the fryer temperature using an immersion thermometer that usually comes with the fryer. The problem is that the fat in which a turkey is fried represents a large volume of flammable liquid which can ignite if overheated, and there is no means of automatically limiting the fat temperature. The fryers are intended for use only outside, never inside a building, yet numerous instances exist where a fryer was close enough to a house for a fryer fire to spread to the house. Because of the propensity for fryers to overheat and ignite if not monitored continuously, fryers are considered defective by design because they lack automatic temperature controls.

Providing an automatic temperature control would make a fryer more expensive, but the hazard of overheating, in terms of both property damage and possible personal injury, is usually considered to outweigh the higher price.

Persistent Water Leaks

Leaks from several bolted joints in a system of plastic water pipes damaged a newly constructed high-rise residential building. Despite the installation contractor having repeatedly retightened the bolts before the incident, the building owner assumed the leaks were caused by faulty installation of the pipes. The owner therefore refused to pay the installation contractor the nearly $300,000 owed under the contract. Consequently, the installation contractor hired a forensic mechanical engineer to determine the true cause.

The pipes were situated on the top floor of the 16-story building and were part of the building's heating and air conditioning system. Water was heated by boilers in winter and cooled by chillers in summer and circulated to fan-and-coil units throughout the building. The failed joints were in the main pipes near the boilers and chillers. The pipes were made of plastic (polypropylene) and manufactured by a large, international corporation purportedly for this type of application. The joints were flange-type, meaning that each end of a pipe featured a large diameter ring (flange). Bolts passed through holes in mating flanges and were secured with nuts. A rubber gasket between the flanges made the joints watertight.

The leak-tightness and reliability of the joints depended on the nuts being properly tightened. Nut tightness is measured by the torque (turning force) applied when the nuts are installed on the bolts. The required torque was specified by the pipe manufacturer and was in fact used by the pipe installation contractor. The contractor used a torque wrench to tighten the nuts, which is an adjustable wrench that automatically applies the selected torque. The torque wrench was brand new.

Certain joints began leaking about eight months after installation. The building owner called the installation contractor who found many nuts loose which they retightened to the pipe manufacturer's torque specifications. The gaskets were also replaced. All other flanged joints were also checked and some nuts were found loose and were retightened.

Over the next seven months, several more joints leaked despite all joint nuts having been checked and retightened as necessary more than 30 times by the installation contractor. In view of the repeated loosening, several remedies were tried, including adding washers and lockwashers beneath the nuts and changing the type of bolts. The chronic loosening nevertheless continued, and one particular leak caused substantial damage.

Ultimately, the forensic mechanical engineer determined the cause by inverting the problem. Rather than concentrating his analysis on the bolts and nuts, the engineer also considered what the nuts were bearing on – the plastic.

The engineer discovered that the temperature of the hot water carried in the pipes in heating season, 150 °F, was the same temperature at which the plastic pipes begin to lose strength and deform (yield) under compression. The engineer also discovered that the gaskets were deforming (losing resilience) because they were the wrong type for the temperature and the specified nut torque, and that the torques specified by the pipe manufacturer were as many as eight times higher than they should have been.

Nuts create compressive force the same way as wedges do. Nut threads are simply spiral wedges between the bolt threads and the surface on which the nut bears. Applying torque to nuts therefore creates tension in the bolts *and* compresses the bearing surface. Like wedges, nuts remain in rotational position (tight) due to friction. The frictional resistance to nut counter-rotation (loosening) depends on the downward (wedging) force imposed on the bolt threads *and* the compressive resistance of the bearing surface.

In a flange joint, leak-tightness is created by gasket compression. Leakage therefore means that compression was lost and/or the gasket deformed. Compression can be lost in two ways: either the nuts counter-rotate (loosen) or the surface upon which the nuts bear yields (deforms) under the compression load of the nuts. Early attempts to diagnose the leakage problem only considered nut counter-rotation to explain why the nuts were loosening. It was not recognized that nuts will also loosen without having counter-rotated if the nuts' bearing surfaces deform, which was ultimately determined to be the problem. Similarly, the gaskets were deforming because the compression to which they were subjected was far greater than they were designed to withstand.

(No one thought to mark the positions of newly retightened nuts in order to later tell if they actually counter-rotated. Had the nuts been marked after tightening, it would have been clear that they had not changed their rotational positions.)

The root causes of the plastic having given way were twofold. First, the pipes were being used for water temperatures that were too high for the plastic material. Proper design requires that a margin of safety – a difference between the failure temperature of a material and its expected operating temperature – be provided so a system will not fail in the face of ordinary temperature fluctuations or if a control malfunctions. In this case, where the normal operating temperature was 150 °F, the pipe material used should have been safe up to at least 200 °F.

Second, per the engineer's calculations, the nut torques specified by the pipe manufacturer were at least eight times higher than they should have been for the plastic material and the type of gaskets. Lower torques would have meant far less load on the plastic, so the plastic and the gaskets would have been less likely to deform and lose resiliency.

Responsibility for the leakage problem and resultant damage lay with two parties – the design engineer who chose a type of pipe whose temperature rating was essentially equal to the hot water temperature in the system, and the pipe manufacturer for specifying incorrect torques. Torque specifications are determined by the manufacturer, who is usually the only party having enough information about their product to make that determination, so it is proper for an installation contractor to use the manufacturer's torque specification in their work.

Upon learning of the engineer's findings, the owner immediately paid the pipe installation contractor and arranged to have the leakage problem corrected permanently.

MANUFACTURING DEFECTS

A device might be properly designed but defective if not properly manufactured. Incorrect assembly, inaccurate machining, loose fasteners, faulty welds, insufficient lubrication and use of unsanctioned materials are among the many kinds of errors that can result in a faulty product. Forensic engineering determines the nature of, and responsibility for, manufacturing defects.

Hubble Telescope Mirror

When it was created, the Hubble Space Telescope probably ranked as one of the most exquisite inventions to date. Its complexity combined with the stresses it experiences in launch, the hostile environment in which it operates, the accuracy with which it must be controlled and the extremely fine tolerances necessary in manufacturing its 8 ft diameter mirror make its scientific success and long life almost miraculous. Yet, when the results of its initial observations were received, it was immediately apparent that something was gravely wrong. The images were blurry.

Investigation by NASA and the contractors determined that the curvature of the mirror was slightly incorrect. The mirror was fabricated using the most advanced, most precise, computer-controlled manufacturing techniques which should have resulted in the exact curvature required for perfectly sharp images, yet the curvature was wrong. The curvature specifications were the first factor to be checked: they were correct. The manufacturing processes were then reviewed and no errors were found in the detailed records. The formula of the glass was checked and compared with samples of the same batch used in the mirror, and it was found to be correct. Launch data was scoured to see if the telescope could have experienced unexpected stresses, but none were indicated. How, then, could the curvature be wrong?

Manufacturing the mirror entailed casting the glass into approximate shape and then grinding the surface to the exact required curvature. Because of the extremely fine tolerances to which the mirror had to be ground, ordinary measurement techniques would not be capable of monitoring how accurately the grinding process matched the specified mathematical curvature. Instead, an optical comparator connected to the computer-controlled grinding machinery was used to assure the ground surface matched the specified curvature. An optical comparator is a device that shines laser light onto a surface and measures distances to an accuracy of less than the wavelength of light. As the grinding proceeds, the computer calculates how closely the curvature corresponds to the specified curvature and automatically adjusts the grinding machinery to correct for any discrepancy. The comparator's laser light is focused by a lens, so the precision of the lens directly affects the precision of the grinding. After the curvature error was discovered, the comparator was tested and its lens was found to be inaccurate. In other words, the telescope mirror was perfectly ground to an imperfect curvature.

When removed from the comparator, the lens itself was found to have been manufactured accurately, but it had been incorrectly installed in the comparator. It transpired that at the comparator manufacturer's facility, the two technicians assigned to assemble the comparator reportedly did not know exactly how the lens would fit and they decided not to ask. Instead, they somehow force-fitted the lens in backward. The stress imposed on the lens distorted its curvature enough to affect the comparator's accuracy.

Further investigation revealed that the comparator had not been tested before being used to grind the telescope mirror, and the telescope mirror had not been tested before the telescope was assembled, strapped to the rocket and launched into orbit. Why not? Cost. Testing the comparator and telescope mirror would reportedly have cost $100,000, which together with other unexpected expenditures, threatened to push the project over its budget, so testing was abrogated. This was a classic case of "penny wise and pound foolish," because it took millions of dollars to develop corrective optics and conduct a manned repair mission to the orbiting telescope to install the optics. Fortunately, the corrective optics performed perfectly and subsequent images from the telescope are perfectly sharp.

Hedge Trimmer Trigger

A homeowner purchased a battery-powered trimmer to trim bushes surrounding his house. The trimmer featured a two-foot-long, reciprocating blade; a housing at the rear containing the motor, battery and trigger switch; and a top handle at the front of the housing that incorporated a second switch. The trigger switch and the front handle switch both had to be depressed for the motor to run. The homeowner was a construction contractor experienced with using tools of all kinds. After charging the battery, he used the trimmer for the first time without a problem. When finished, he carried the trimmer by the rear handle – the blade hanging downward – to his storage shed and placed the trimmer on a table. Later, to move the trimmer onto a shelf, he lifted the trimmer by the front handle, which depressed the second switch, and the trimmer unexpectedly turned on. He had not pulled the trigger. The blade seemed to be swinging toward his leg, so he fended it off with his free hand, which was lacerated by the blade. After obtaining medical care, he returned and examined the trimmer and discovered that the trigger was stuck in the ON position. He photographed and videotaped the trigger in this position and then tested the trigger, confirming that it tended to stick in the ON position.

Forensic investigation confirmed the trigger's tendency to stick ON and that the front handle switch operated normally. Internal examination of the handle revealed that the trigger consisted of plastic bar. When the trigger was pulled, the end of the bar slid past and depressed the actuation button of a small microswitch. The bar was spring-loaded so that at rest, the actuation button was not depressed. The face of the bar that slid past the microswitch was found to be slightly rough, which, although the face was lubricated, was enough to cause the trigger to sometimes fail to return to the OFF position when released.

It was clear that the trigger was defectively manufactured such that its face was not smooth.

INSTALLATION DEFECTS

A device might be properly designed and manufactured but rendered hazardous by being improperly installed. Improper installation of a chimney pipe can lead to a fire. Improper installation of insulation can allow a pipe to freeze. Improper installation of a boiler without proper pressure relief capability can lead to an explosion if a control malfunctions. Improper installation of pipes without pressure testing can lead to a leak.

Fuel Oil Leak

A heating contractor replaced an old, worn-out, oil-fired furnace. Oil was supplied from a conventional tank in the basement via a standard 3/8″ diameter, soft copper line. The line had been connected to the old oil burner using a compression fitting. The contractor reused the existing line and compression fitting. Not long thereafter, oil was discovered on the basement floor and oil fumes permeated the house which necessitated costly remediation.

Forensic mechanical investigation discovered that the source of leakage was the oil line connection to the oil burner. The connection was tight but was internally distorted and unable to form a leak-tight seal. Because oil leakage is hazardous and compression fittings are notoriously unreliable, especially when

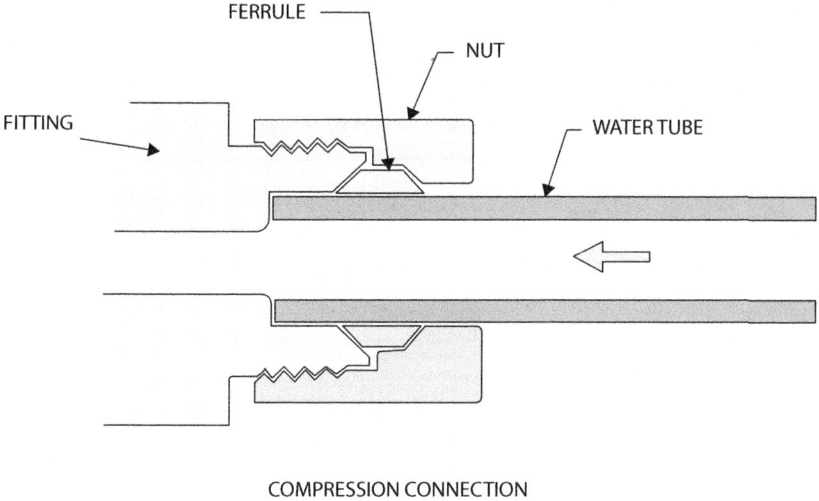

FIGURE 2.1 Compression fitting.

re-used, compression fittings should never be used on oil lines. Flare fittings should be used instead.

A compression fitting uses a small brass ring (ferrule) near the end of the copper line (see Figure 2.1). The leading and trailing edges of the ferrule are beveled (like a man's wedding ring). Behind the ferrule is an internally threaded compression nut with a beveled interior shoulder. When the nut is tightened onto the threads of a mating fitting, in this case on the burner's oil pump, its shoulder wedges the ferrule tightly into the small gap between the outside of the line and the inside of the mating fitting and simultaneously compresses the ferrule into the surface of copper line. This forms a leak-tight seal at the oil pump and a mechanical connection between the line and the pump.

The problem with reusing an existing compression fitting is that the wedging force created in retightening the nut can deform the copper line enough to prevent the formation of a leak-tight seal. Usually, immediate testing by opening the outlet valve on the oil tank reveals leakage, but sometimes leakage does not occur until the next time the tank is filled, which increases the pressure in the line. When the leak is discovered, further tightening the compression nut usually only exacerbates the leak by increasing the deformation of the line. Therefore, proper repair requires replacing the compression fitting. Replacing the compression fitting means shutting off the oil tank, detaching the fitting, cutting off a couple of inches from the end of the line containing

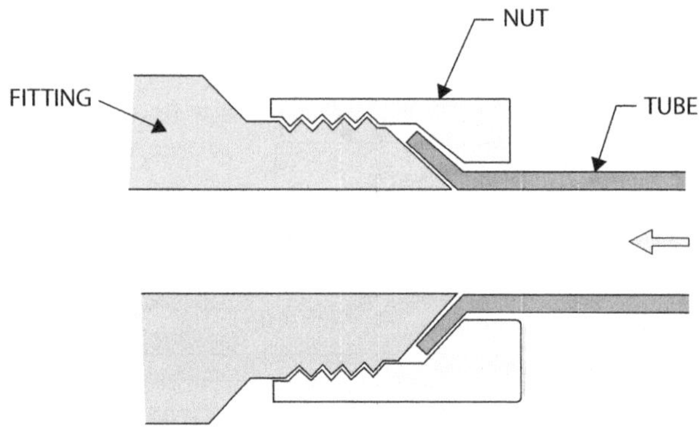

FIGURE 2.2 Flare connection.

the compression fitting and installing a new fitting on the fresh, undeformed end of the line.

The unreliability problem with compression fittings is avoided by using flare fittings instead, which in fact are mandated for oil lines (see Figure 2.2). In a flare fitting, instead of a ferrule, the end of the copper line is expanded outward into a short cone. A small, portable flaring tool is used for this purpose in order to create a perfect flare. A flare nut on the line behind the flared end has an internal shoulder beveled at the same angle as the flare. The flare nut connects to a mating flare fitting (such as on an oil burner) whose end is also beveled at the same flare angle. In this way, the flare cone is sandwiched between the flare nut and the flare fitting. Tightening the nut does not deform the line. Any leakage is easily corrected by simply tightening the flare nut a bit more. A flare connection forms a solid, leak-proof, mechanical attachment between the line and the device to which the line connects, such as an oil burner.

Thus, the use (or reuse) of a compression fitting represents defective installation of an oil line.

Aircraft Panel Failure

While a commercial aircraft was in flight, a large panel on the side suddenly flew off, instantly depressurizing the plane. Fortunately, no one was injured and the plane was able to land safely. The panel, known as a "door plug," was a surrogate used when the owner of a plane does not want a door where the plane's structure has been designed for one. The plug was supposed to be secured to the structure by several bolts which were found to have been omitted when the plane was manufactured. The defect proved extremely costly to the airline and the manufacturer because all other aircraft of the same model had to be taken out of service and their door plugs inspected for integrity.

Shower Leak

A wet spot was discovered in the first-floor ceiling of a house at a point below the stall shower in the master bathroom. The master bathroom had been renovated 10 years ago. The new stall shower was all tiled, including the shower doorway. To repair the leak, a plumber opened the ceiling and discovered heavy rot on the wood structure. The leak was small but had clearly existed for a long time and only recently created the wet spot. Based on the damage pattern, the plumber concluded that the leak site corresponded with the doorway of the stall shower.

The plumber ran water in the shower to test the pipes, faucets, shower head and floor pan and found no leaks. The forensic mechanical engineer retained by the insurance company located the leak on top of the sill of the shower's doorway at a small crack between a tile and grout at one end of the sill.

Although cracks may develop between tiles and grout due to thermal expansion and contraction, proper installation of tile should normally not compromise water-tightness. In this case, the tiling was installed incorrectly. The tiles on the sides of the doorway had been installed before the tiles on the sill. In other words, the sill tiles had been installed between the bottom ends of the side tiles. Thus, the leftmost and rightmost sill tiles butted into the vertical side tiles. The crack was in the leftmost joint. The significance is that a crack in this joint allowed water to seep downward into the structure. By contrast, the sill should have been tiled first and the side tiles installed second, so the lower ends of the side tiles would have butted into the horizontal surfaces of the sill tiles, thus creating a barrier to water penetration.

SERVICE DEFECTS

Sometimes a perfectly good appliance or machine fails and requires service, and then, after service, a serious malfunction occurs. Sometimes the malfunction was the result of defective service.

Boiler Explosion

A retired couple owned a modest, two-story home with a full, unfinished basement located on a sloping rural property. The home was heated by a gas-fired hot water boiler in the basement. One winter day, the boiler failed and they called their regular heating contractor for service. The contractor sent a young, inexperienced technician who decided the problem was electrical in nature, so he replaced all three of the boiler's principal electrical controls, including the aquastat. The aquastat is the master control that turns the boiler and the water circulation pump on and off in response to the house thermostat. The boiler began running, and he left. Fifteen minutes later, the boiler exploded. The explosion blew the boiler off its foundation and blew out the exterior wall of the basement. The homeowners were upstairs and were injured by the shock.

After the fire department shut off the gas line and made sure the basement was safe, the homeowners contacted their insurance company and their lawyer. The insurance company agreed that the lawyer would pursue a possible action for damages and recovery of the damage claim that the insurance company paid. A forensic mechanical engineer is retained to examine the evidence and gather information to establish the cause and responsibility for the incident.

The boiler and all controls were examined and tested wherever possible. Tracing the wiring among the controls was especially difficult because the wires had been cut in order to enable the boiler to be removed from the basement. Ultimately, it was found that the boiler's new aquastat had been miswired when it was installed by the technician shortly before the explosion. The miswiring prevented the boiler from ever turning off and prevented the circulation pump from running. As a result, the boiler simply grew hotter and hotter until its back wall got red-hot, deformed under steam pressure and blew out.

(An ironic twist is that the young technician was actually the homeowner's son in his first employment after completing his training as a heating system repairman.)

Persistent Soot Emission From an Oil-Fired Furnace

A modest-sized condominium was heated by an oil-fired warm air furnace. The owner began noticing black dust on household surfaces, and suspecting it was soot, called a heating system service contractor. The contractor checked the furnace and oil burner and found nothing physically wrong, such as a crack in the heat exchanger, which might allow sooty exhaust to escape into the household airstream. He checked the nozzle, which was relatively new and clean. He checked the oil filter, which was clean and clear. He tested the combustion parameters, including draft (chimney suction) in the combustion chamber and flue pipe, and the oil pressure. He checked the furnace air filter and found it soot-blackened as expected. He checked the air circulation fan, finding it normal. He made a tiny adjustment to the draft regulator setting, but otherwise could not explain the sooting problem.

The problem persisted. Eventually, the owner had to hire a cleaning contractor to remediate the soot and filed an insurance claim for the damage. The insurance company retained a forensic mechanical engineer to determine the cause. The engineer rechecked the furnace's combustion parameters, which seemed normal, and then disassembled the oil burner, which revealed a small, critical anomaly that solved the mystery.

An oil burner consists of a body and a blast tube, which is the pipe that carries combustion air into the furnace's combustion chamber. Within the blast tube is the pipe that carries oil to the nozzle. As discussed more fully in Chapter 7, proper mixing of air and oil vapor is essential for soot-free combustion. Air-vapor mixing is created by a vaned plate at the far end of the blast tube. The nozzle is situated at the central hole of the vaned plate. Air passing through the vaned plate is swirled to intermix with the nozzle spray (see Figure 2.3).

The exact distance of the nozzle from the vaned plate, called the "Z-distance," is critical for proper intermixing. In this case, the distance was slightly incorrect, which resulted in high-frequency flame pulsations. Each

FIGURE 2.3 Vaned plate at end of blast tube, with nozzle and ignition electrodes visible within central hole.

pulsation consisted of a high-pressure phase and a low-pressure phase, both extremely brief. Each high-pressure phase propelled a tiny amount of soot-laden exhaust into the household environment. The pulsation frequency was too high to be audibly or visually noticeable or to register on a draft gauge which is designed to measure average draft. The serviceman never checked the Z-distance because once set, the Z-distance should never vary, but in fact was wrong. He should have checked everything. Resetting the Z-distance eliminated the pulsations and resolved the soot emission problem.

IRRELEVANT DEFECTS

Sometimes an investigation will reveal a defect in a product or process that is real but did not contribute to the incident at hand. In this situation, the hazard is that the client may try to assert that the defect is relevant even though it is not. The engineer's duty is to convince the client that a legal action would not likely prevail. At most, the client might hope to negotiate a settlement, but even that could be costly as well as fruitless.

Multiple Furnace Defects, All Non-causal

A modest-sized house in a suburban neighborhood froze and suffered water damage while the owners were in Florida for the winter. The incident was discovered by their daughter during a regular weekly check. The house was heated by a gas-fired warm air furnace, which had failed, so water pipes froze and burst.

The furnace was a high-efficiency type that had been installed nine years earlier and had never failed before. The furnace drew combustion air directly from outside and vented exhaust directly out of the house. Four days after the incident, a service contractor (different from the contractor who installed the furnace) was able to restart the furnace, but 22 days later, the furnace failed again. Comprehensive troubleshooting revealed six problems, including that the vent pipe was undersized, the gas supply pipe lacked a drip leg to trap foreign particles, the house thermostat was faulty, the furnace's electronic control board was faulty and the outside air temperature sensor was faulty. The undersized vent pipe and lack of a drip leg were installation defects. The other defects were probably consequences of the water damage.

A forensic mechanical engineer was retained to determine if the installation defects – the inadequate vent pipe and lack of drip leg – were causally significant to the incident. Analysis of the heating system and circumstances confirmed that the vent pipe was undersized (2″ instead of 3″ size) in violation of the furnace manufacturer's specifications. A drip leg is a precaution against foreign matter in the gas stream being conveyed to the gas controls, but in this case, no evidence was found that the furnace failed because foreign matter affected the function of the gas controls. The actual cause was the failure of the electric valve on the gas line, which had nothing to do with any of the defects. The valve failed electrically, not mechanically.

FAILURE PROBABILITY

A failure, fire, or accident may be caused by simple wearout, which would not be considered a defect unless the wearout were significantly premature. Products have a certain expected average life, which might be, for instance, 20 years for a heating system, 13 years for a car, 12 years for a water heater, 6 years for a hair dryer, etc. Some failures will occur sooner and some will occur later than the average. A product would only be considered defective if it fails substantially sooner than its expected average life. For instance, if a hair dryer fails after only one year, which is far sooner than the six-year average lifetime of that type of appliance, the failure would be reasonably attributable to a manufacturing defect.

Peculiar Failures

3

INTERESTING ODDITIES

Products fail in a variety of ways. A product might not turn on or might simply stop during use. It might work only in certain circumstances but not in others, such as in dry weather but not in humid weather. A heating appliance might suddenly produce less heat than normal. Adhesive tape might gradually become overly sticky so it cannot be pulled off the roll. A television might suddenly become unable to receive certain kinds of programming. A water heater might develop a leak. These kinds of failures are not unusual and the causes are usually not difficult to identify. However, some failures are unusual or uncommon, and therefore interesting.

Tile Shower Leak[1]

A couple of years after a stall shower was installed when a second-floor bathroom was renovated, water leakage was observed on the first-floor ceiling below. Fresh leakage never occurred when the shower was not in use and did not occur every time it was. A leaking pipe or valve was ruled out because either of these causes would occur every time the shower was used. The shower walls were tiled, and a small crack was found between a tile and the adjacent grout. But why didn't the joint leak every time the shower was used?

A forensic engineer solved the mystery by having the homeowners keep a log of shower use, showing who showered, the date and whether leakage occurred. The log showed that leakage occurred only when the husband showered, never when the wife did. The husband was six feet tall and the wife was 5′4″ tall, so water splashed off each at different heights. Only splashes from the husband were high enough to reach the leaking joint.

DOI: 10.1201/9781003684213-3

Boiler Moaning Sound

Homeowners decided to have an addition built onto their house. The addition, which had no basement, was large enough to require its own heating system, so a gas-fired, tankless hot water boiler was installed on an outside wall to supply radiant heating pipes in the concrete floor. The boiler drew combustion air and expelled exhaust through the wall. The new heating system operated properly in terms of heat production, but it generated a moaning sound. A heating system contractor was called and found nothing operationally wrong, but could not explain the sound. He called the boiler manufacturer, who also was unable to provide a remedy. A forensic mechanical engineer eventually pinpointed the problem, which was resonance in the air intake pipe.

Resonance is a phenomenon where an oscillation or vibration in one part of a device or system stimulates sympathetic oscillation or vibration of another part of the device or system. In some situations, resonance is beneficial, while in others it may be annoying or even catastrophic. Beneficial resonance creates the sound in a musical instrument, like vibration from air blown at the edge of a reed in an oboe, vibration in the player's lips creates sound in a trumpet, and vibration from air flowing past the lower edge of an organ pipe creates sound in the pipe.

Every object has certain natural frequencies of vibration that govern the nature of the sound produced when the object is struck or subjected to vibration. The natural vibration frequency depends on the size, material and shape of the object, for instance, the size and shape of a bell, the length and diameter of a trombone pipe or an organ pipe, the length of a playground swing, or the design of a bridge. The key is whether an imposed vibration or oscillation matches the natural frequency of the object. If the frequencies match, the object's vibration will amplify; if not, the imposed vibration will die out. If a playground swing is pushed at the top of its swing, the swing will progressively amplify. If the swing is pushed anywhere else is its arc, the movement will be dampened. If a suspension bridge is impacted at a certain frequency such as by a parade of people marching in step or if wind at a certain speed excites a bridge's natural frequency, the bridge may begin to oscillate or swing ever more violently to the point of collapse.[2]

Just as with an organ pipe, the moan from the boiler was a resonance created by the blade frequency of the boiler's air intake fan drawing air at a certain velocity through a pipe of a particular size, length, material and configuration (a sequence of elbows). The remedy was to alter these conditions. The contractor removed a short section of the intake pipe, which eliminated the moan, but only when the boiler was operating at its high firing rate, when heat demand was maximum. The moan persisted when the boiler was operating in its low firing rate in mild weather. The remedy was further alteration of the intake pipe, the specifics of which required experimentation.

Water Heater Leak

A modest-sized, two-story home was equipped with an 11-year-old gas-fired water heater located in the basement. The water heater had always functioned normally and had never been serviced. One winter, the homeowners went south for three months. When they entered the house upon returning, everything seemed normal. They took their suitcases upstairs to unpack, then the husband took the suitcases to the basement for storage. He opened the basement door, stepped down onto the first step, and his foot was immersed up to his ankle. The entire basement was full of water up to the ceiling.

The fire department pumped out the basement and shut off the main water valve. Subsequent investigation revealed that the water came from the water heater. No other leaks were found in any pipes in the basement and the rest of the house was dry. Groundwater had not infiltrated through the foundation. A forensic mechanical engineer examined the water heater, removed the sheet metal jacket and insulation and found a corrosion hole in the pressure tank less than three-sixteenths of an inch in diameter. The amazing thing was how much water leaked through such a small hole in three months – thousands upon thousands of gallons.

Repeated Heating System Failures

A couple whose primary home was in the Northeast owned a second home in Florida, where they went for the winter. One spring, when they returned, they discovered substantial water damage from a burst pipe. The heating system was not working. The pipe had burst from freezing. Their heating system contractor discovered the batteries in the programmable thermostat had failed. He replaced the batteries and the heating system operated normally. The homeowner had not installed fresh batteries before leaving for the winter.

The water damage was costly and took several months to fully remediate.

The following winter, the couple went to Florida again as usual. When they returned in the spring, they found the house again flooded from a burst pipe and the heating system again not working. Their heating system contractor found that the thermostat batteries had failed. Despite the previous arduous experience, the homeowner again had neglected to install fresh batteries before leaving for the winter.

Notes

1 This is a different shower than the one in the example under Installation Defects above.

2 For instance, the Tacoma Narrows bridge that was constructed in July 1940 and collapsed in high winds in November 1940.

Machine Accidents 4

INJURIES IN AN INDUSTRIAL ENVIRONMENT

Despite modern safety systems and procedures, industrial accidents still happen. A forensic mechanical engineer is often hired to determine how an accident happened, whether a defect was involved, and who, if anyone, is responsible. Although similar in many ways to any other forensic engineering investigation, a machine accident investigation has to concentrate more on the machine. How does the machine work? Did the machine malfunction? Did a component fail, and if so, was it insufficiently strong, improperly manufactured, fatigued, corroded, or simply removed for some reason? Did the design of the machine compel the operator to take a dangerous action? Was the hazard perceptible or covert? Was the machine being used for its intended purpose? Was the misuse foreseeable? Had the machine been modified in a manner that created the hazard? Was the site of the accident mechanically or electrically unguarded or was the guarding improper? Was the guard removed, and if so, why? Were the machine's controls easy to use, difficult to use, properly or improperly located? Were the controls adequate, guarded or shielded? Were all necessary warning labels present or were some omitted, worn off or painted over, incomprehensible or incomplete? Given the level of experience of the injured worker, would a warning label have made any difference?

If a machine malfunction was involved, was the malfunction constant or intermittent? Simple testing often cannot replicate an intermittent malfunction, in which case the malfunction can only be substantiated by examining a part whose malfunction would produce the reported action.

Laboratory analysis might be necessary, such as chemical analysis of a lubricant, metallurgical analysis of the mode of failure or fracture, or microscopic examination to reveal possible contaminants.

Every inspection must be documented in the form of notes and photographs, and sometimes videotaping of machine motion or the action of the injured person if those actions are complex and difficult to capture with still

DOI: 10.1201/9781003684213-4

photography. The entire vicinity should be photographed in case a question arises later after the machine or its environment has been modified.

Two examples are relevant. The first, although not a machine accident, is the "Faulty Chimney Pipe Installation" case in the Arson chapter. The second is the following.

Heating System Scalding

A huge commercial greenhouse for growing plants wholesale was heated by hot water generated by two boilers and piped to fan-and-coil units throughout the building. The piping system contained pumps and control valves at strategic locations. One day, a maintenance technician was assigned to replace a faulty pump. The task required closing the valves on all pipes associated with that pump, draining the pump and installing a new one. The maintenance man closed three valves near the pump and proceeded to remove the bolts holding the pump to the piping. With all bolts removed, he banged on the pump to loosen it, and when the pump cam free, he was sprayed with scalding hot water. It turned out that he had failed to close a fourth associated valve. The valves had not been marked as to which pump they served, nor had the pump been marked with how many valves it was associated with. Some pumps had three associated valves and some had four.

Forensic engineering investigation confirmed the circumstances of the incident, but a question was later raised about the identity of the valves in question. Fortunately, the issue was resolved by reviewing the photographs of the area taken by the forensic engineer.

Combustion Explosions

5

IDENTIFYING AND QUANTIFYING THE FUEL

The first objective in an explosion investigation is to identify the fuel involved. Was it natural gas in a house explosion, fuel oil in a boiler explosion, chemical vapor in a textile plant, or dust in a granary? The site and circumstances can narrow the possibilities, but confirmation may require forensic analysis. For a given possible fuel, could enough have accumulated to cause an explosion of the apparent magnitude? The following discussion applies to combustion explosions, not bombs, steam explosions or pressure vessel ruptures.

EXPLOSION MECHANICS

An explosion causes damage by exerting force on surrounding objects. An explosion in a house can blow out windows and walls. An explosion in a manufacturing plant can disrupt machinery. An explosion in a furnace can blow the furnace apart. Combustion produces force by heating the air in which the explosion occurs. Heating causes air to expand. If the expansion is constrained, the expansion produces force on the constraints, such as the walls within which the explosion occurs. Sudden combustion produces sudden expansion, the shock of which can fracture structural members.

Heat-induced expansion of air increases the air pressure in exact proportion. The pressure increase, known as "overpressure," is imposed on all objects in the vicinity. Overpressure is measured in pounds per square inch (psi). For instance, an 8″ × 12″ window exposed to a 2 psi overpressure (such as from an explosion) will experience 8″ × 12″ × 2 = 192 lb of total force. An ordinary window would not likely be able to withstand that much force and would blow out.

DOI: 10.1201/9781003684213-5

EXPLOSION FUEL AND QUANTITY

Determining the type and quantity of fuel involved in a combustion explosion begins with assessing the damage and measuring the space in which the explosion occurred. The space size plus the magnitude of the damage leads to an estimate of the overpressure involved, and the overpressure leads to the quantity of fuel. The final step is to determine how the necessary quantity of each possible type of fuel could have accumulated – a pipe leak, a gas tank rupture, a flammable liquid spill, etc. The result of this analysis leads to confirmation of the type of fuel and the nature of the explosion incident, from which conclusions can be drawn about responsibility.

Although means of detailed calculation exist for relating the degree of damage to the magnitude of an explosion,[1] the following table describes the approximate relationship between damage and overpressure, which is sufficient for many explosion investigations.[2]

EXPLOSION EFFECTS	
OVERPRESSURE	*DAMAGE*
0.5–1 psi	Windows shatter, cinderblock walls collapse.
1–2 psi	Asbestos siding shatters, corrugated steel panels buckle.
2–3 psi	Non-reinforced concrete walls 8–12″ thick shatter.
2.5 psi	Moderate damage to light steel-frame industrial buildings.
3.5 psi	Human eardrums burst.[3]
4 psi	Moderate damage to steel-frame industrial buildings and load-bearing brick walls.
5.5 psi	Moderate damage to steel-frame buildings and reinforced concrete multi-story buildings; severe damage to load-bearing brick buildings.

Calculations are based on the following two key physical constants[4]:

- One cubic foot of air weighs 0.075 lb at ordinary room temperature.
- It takes 0.24 BTU of heat to raise the temperature of one pound of air one degree Fahrenheit.

The following table shows the heat produced by burning common fuels.

HEAT OF COMBUSTION OF COMMON FUELS	
Acetylene	1,448 BTU/cu.ft.
Coal (bituminous)	Approximately 14,300 BTU/lb
Ethyl alcohol	1,575 BTU/cu.ft. vapor
100 octane gasoline	Approximately 6,000 BTU/cu.ft. vapor
Natural gas	Approximately 1,050 BTU/cu.ft.
Propane	2,516 BTU/cu.ft. vapor

For example, how much propane would it take to blow out the windows of a 10 ft. × 12 ft. × 8 ft high kitchen? The volume of the kitchen is 10′ × 12′ × 8′ = 960 cu.ft. The weight of air in the kitchen is 960 cu.ft. × 0.075 lb/cu.ft. = 72 lb. From the explosion effects table, it takes approximately 1 psi overpressure to blow out a window. Overpressure calculations utilize absolute pressures (psia), so with normal atmospheric pressure being 14.7 psia, 1 psi overpressure is 15.7 psia. The ratio of overpressure to ordinary pressure is 15.7/14.7 = 1.068. (This can be thought of as a 6.8 percent increase.)

The temperature increase necessary to produce a given overpressure is exactly proportional to the pressure increase, in this case, 6.8 percent. The temperature increase must also be measured on an absolute scale in which zero absolute is 460 degrees below zero Fahrenheit. Assuming the kitchen temperature was 70 °F, the absolute temperature was 460 + 70 = 530 degrees absolute. Using the 1.068 ratio of overpressure to ordinary pressure, the temperature required to produce a 1 psi overpressure is 1.068 × 530 deg A = 566 deg A, or 36 °F.

To heat 72 lb of air by 36 °F requires 0.24 BTU/lb = 622 BTU.

Using the heat of combustion table, the quantity of propane vapor to generate 622 BTU of heat is 622/2,516 BTU/cu.ft. = 0.247 cu.ft. In other words, it would take an explosion involving less than a quarter cubic foot of propane vapor to blow out the windows of a 10′ × 12′ × 8′ kitchen.

Kitchen Explosion

A young couple rented a two-story frame house. The kitchen on one side of the rear featured a gas range, above which was a wooden shelf on which sat a small, 6 ounce can of red spray paint.

One afternoon, an explosion blew out all the windows in the house and split off the exterior wall of the kitchen a couple of inches. The wife was sleeping on the sofa at the time of the explosion.

In the kitchen, the fire marshal found the spray can split open, obviously from extreme internal pressure. Red paint was splattered on the wall behind the shelf and the shelf itself was split. The fire marshal decided that the small amount of propellant – propane – in the can would have been insufficient to cause so large an explosion, so he concluded that the explosion was caused by someone cooking drugs on the stove. However, no evidence or drug paraphernalia was found to support that opinion. Based on the fire marshal's conclusion, the couple's renter insurance denied coverage based on the policy's illegal action exclusion.

The kitchen was the same size as in the previous example, 8′ × 10′ × 8′ high, and contained 72 lb of air. Based on the shattered windows and the wall having been shifted outward, the explosion overpressure was taken as 1.5 psi. This overpressure is an increase of 16.2 psi/14.7 psi = 1.102, or 10.2 percent. The absolute temperature increase corresponding to 10.2 percent is 1.102 × (460 + 70) = 54 °F. Increasing the temperature of 72 lb of air by 54 °F requires 0.24 × 72 × 54 = 933 BTU of heat. Propane combustion generates 2,516 BTU per cu.ft. of vapor, so 933 of heat corresponds to 933/2,516 = 0.37 cu.ft. of propane vapor. From standard tables, one ounce of liquid propane transforms into 0.28 cu.ft. of vapor. Therefore, the explosion was fueled by 0.37/0.28 = 1.3 ounces of liquid propane. In spray cans, pressure keeps most of the propellant in liquid form. The quantity of propellant in the red spray can is unknown, but 1.3 ounces is only about one-fifth of the 6-ounce can. Thus, the small amount of propane necessary to fuel the explosion comports with the can propellant, which was the fuel for the explosion, thus contradicting the fire marshal's seat-of-the-pants opinion that ruled out the spray can.

In fact, an interview with the woman revealed that shortly before the explosion, she had turned on one of the stove's rear burners in preparation for putting a pot on to heat. However, the phone rang in the next room before she was able to put the pot on. While talking on the phone, she forgot the burner was on, and when she hung up, she went to take a nap. Meanwhile, heat from the burner warmed the spray can, thereby raising the pressure until the can exploded, releasing the propellant, which was then ignited by the burner to produce the explosion.

Based on the forensic calculations, the insurance company paid the renters' claim.

Cooking Torch Explosion

A fireball suddenly erupted from a small kitchen torch after several unsuccessful attempts by the user to light it. The user was burned about the face and torso. The user asserted that she had not actuated the ignition button when the fireball occurred.

The torch is the kind commonly used in cooking to superficially darken foods during preparation. The torch is fueled by butane which is contained in a plastic tank inside the handle. The torch head, located at the top of the handle, includes the flame nozzle and sparker (igniter) at one end and a pushbutton at the other end which triggers a spark. The tank is refilled via a port at the bottom of the handle. Refilling involves turning the torch upside-down and inserting the refill pipe from a small butane tank into the port. Liquid butane then flows by gravity into the torch tank and displaced butane vapor escapes into the air. For that reason, the nozzle/tank connection is deliberately not tight.

The torch had been previously used successfully by the user's fiance. A few weeks later, the user tried to use it, but the torch would not light. The user read the operating instructions, then refilled the torch. The torch still would not light. She handed the torch to her fiance, who refilled it again, but again it would not light. A minute or two later, the user took the torch back and tried again to light it, but was unsuccessful. Moments later, while she was holding the torch upside-down, a fireball suddenly erupted from the bottom of the torch, which produced substantial burn injuries.

Examination showed no external or internal heat damage to the torch. Even the delicate plastic rim around the base, which enables the torch to stand upright, was in normal condition. Inside the torch, the tank is thin-walled plastic, the tube that carries butane from the tank to the nozzle is slender plastic and fine insulated wires run from the pushbutton through the head to the sparker. These components are all highly vulnerable to heat damage, yet none of them showed any heat damage at all.

The challenge was to reconstruct what happened. The fireball was obviously fueled by butane, but where did the butane come from and what ignited it? The analysis was based on the physical properties of butane, two of which are important.

First, like any flammable liquid, only fuel vapor burns; liquid fuel does not. Butane vapor must be mixed with air in a certain ratio range in order to be combustible (1.9–8.6 percent butane in air).

Second, butane vapor is twice as dense as air, so butane vapor flows downward when released.

A spark, an open flame or other hot source is required to ignite butane vapor. The subject torch is ignited by a spark adjacent to the nozzle, but the user denies having pressed the ignition button when the fireball occurred. The possibility was considered that overly dense, unburned butane vapor had filled the head during the last unsuccessful ignition attempt, so a small flame had occurred where the butane vapor was in contact with air at the front of the head. The flame was postulated to have spread slowly back through the head and down through the handle, expanding from the flame heat to erupt as a fireball from the handle and the time of the explosion. This hypothesis fails due to the absence of heat damage inside the torch.

The other possibility is that butane escaping from the torch during refilling flowed downward and reached an external source of ignition from which flames spread back upstream to the torch to create a fireball. In other words, the fireball *seemed* to erupt from the torch, but was actually just *situated at* the torch. In fact, an older type of refrigerator was located 6–8 feet from where the user was standing. Refrigerator compressors are situated on the bottom, a compressor is powered through a relay and a spark occurs at a relay's electrical contacts when a relay is actuated by a refrigerator's thermostat.

Thus, with ignition inside the torch being conclusively ruled out for lack of heat damage, it was concluded that the fireball was ignited externally. Liability was concluded to rest with the torch manufacturer for lack of sufficient warning about the need to refill a torch in a well-ventilated space and to allow a refilled torch to rest for many minutes in order to allow butane vapor to dissipate before attempting to light the torch.

Notes

1 See for instance National Fire Protection Association Publication No. 68, *Guide for Venting Deflagrations.*

2 "Fire & Arson Investigator," *Journal of the International Association of Arson Investigators*, December 1985.

3 Eardrums are the first part of the human body to be injured by an explosion.

4 R. H. Perry, *Engineering Manual*, 3rd ed., McGraw-Hill, 1976.

Combustion Venting

6

A MATTER OF BUOYANCY

The modern world is extremely energy-intensive. The most useful energy is in the form of heat. Heat is used directly and indirectly to produce electricity, which is the most flexible kind of usable energy. Most heat is derived from combustion because combustion is the simplest source. Most combustion creates toxic exhaust.

EXHAUST

Combustion is the rapid oxidation of a hydrocarbon fuel. A hydrocarbon is a compound composed of hydrogen and carbon, such as oil, natural gas, propane and wood. Each component of a hydrocarbon oxidizes separately. Hydrogen oxidizes to hydrogen dioxide (H_2O) - water. Carbon oxidizes to carbon dioxide (CO_2), carbon monoxide (CO) and usually a small amount of unburned carbon (C). Water is harmless, but the carbon compounds are not.

Carbon dioxide is an asphyxiant, not toxic but not breathable. In contact with water, carbon dioxide forms carbonic acid, a weak acid that attacks mineral structures like limestone, seashells and corals.

Carbon monoxide is toxic. When breathed, blood hemoglobin has 200 times greater affinity for carbon monoxide than for oxygen, so carbon monoxide starves the body of oxygen.

Unburned carbon, called soot, takes the form of tiny particles that can clog the lungs. Soot is notably pervasive and difficult to clean off surfaces.

Combustion can be unconfined, such as in a campfire, wildfire or oil spill fire, or semi-confined, as in a residential fireplace or a burning building, or confined such as in an industrial furnace, boiler or woodstove.[1] Because the products of combustion – exhaust – are dangerous, exhaust from residential

DOI: 10.1201/9781003684213-6

and industrial equipment must be conveyed out of the premises, a process called venting.

EXHAUST VENTING

Exhaust can be vented naturally via a chimney or fan-forced via a pipe. Chimneys work because hot exhaust is lighter than outside air and thus rises by convection. Hot air balloons rise because the air inside is warmer and less dense than the surrounding air. The cooler the outside air, the stronger the upflow through a chimney. The strength of chimney upflow or flow through a pipe is called *draft*. Draft strength affects combustion efficiency in a heat-producing appliance and a weak draft can cause exhaust to escape into the premises, a dangerous situation.

Because the draft is so important and exhaust is noxious, venting equipment must be sturdy enough to remain permanently intact. Also, because exhaust is typically very hot, venting equipment must be fireproof, insulated like masonry chimneys and double-wall metal chimney pipes and spaced a safe distance away from combustibles.

A chimney usually should have a cap to prevent rainwater entry, but more importantly, to prevent wind from suppressing rising exhaust. Besides forcing exhaust to escape elsewhere like into the premises, obstructed upflow can also cause the chimney or chimney pipe walls to overheat and ignite nearby combustibles.

Fire inside a chimney and chimney pipe is a major hazard, especially with oil-fired equipment whose exhaust tends to contain soot (carbon), which is combustible. A chimney fire consists of burning soot on the inside walls. Burning soot on the walls behaves like an upside-down solid-fuel rocket. An immense flame will shoot out of the top of the chimney, and extreme temperatures inside the chimney can melt masonry mortar or cause a metal chimney pipe to glow red-hot. Averting a chimney fire requires assiduous maintenance of oil-burning equipment to minimize soot generation, plus periodic cleaning of chimneys and chimney pipes.

BUOYANCY

Chimneys function due to buoyancy. Buoyancy, where one material rises above another, happens because one material is less dense than another.

(Density is weight per standard volume, like pounds per cubic foot or grams per cubic centimeter.)

Rocks sink in lakes because rocks are denser than water. Life vests float in water because they are less dense than water. Bubbles rise in soda water. Warm air rises above cooler air. Hot air balloons rise. Helium balloons rise. Water balloons sink. Ships float. Submarines float or sink. Airplanes and birds sink – they only elevate when they move, not because they are inherently buoyant.

Warm air buoyancy strongly affects fires. Flames rise because flames heat the surrounding air. In a building, if a window is open or a gap exists in the roof, the flames will flow toward the breach because the outside air is cooler and denser than the fire-heated air.

CHIMNEY EFFECT

The flow toward the breach is called the *chimney effect* because a breach acts like a chimney. The chimney effect can produce a burn pattern that leads back to the point of origin, which helps a fire investigation. However, the chimney effect can also confound a fire investigation by causing the most severe damage to be at the breach rather than where the fire started. For instance, if a fire starts in an attic that has a chimney pipe running up through it, the pipe's roof pass-through can be a breach because the pass-through may be less well sealed than the rest of the roof. Because of the chimney effect, the pass-through area may suffer the most severe damage and falling embers from the pass-through area can obliterate the fire's point of origin. Hence, fire investigators must be conscious of the chimney effect when gathering evidence at a fire scene.

An example illustrating the significance of the chimney effect is found in Chapter 8, "Arson."

Note

1 Liquids heated in appliances called "boilers" may actually boil, like water for steam, or may just produce hot liquid.

Ignition of Mixed Flammable Liquids

7

IGNITION TEMPERATURE IS NOT JUST A SIMPLE AVERAGE

Fire investigators are sometimes called upon to determine the cause of fires involving mixtures of flammable liquids. A very hot ignition source like a spark or open flame will ignite any mixture of flammable liquids, but the challenge is to determine how a mixture could be ignited by a hot object whose temperature is considerably lower than a spark or flame. The answer depends on whether the mixture of vapors coming from the liquid was ignitable at the hot object's temperature.

Only flammable vapor burns; flammable liquid does not. ("Flammable liquid" actually means a liquid whose vapor is flammable.) Vapor naturally evaporates from a liquid at a rate that depends on temperature. In evaporation, randomly moving molecules of a liquid jump up into the air. In condensation, randomly moving vapor molecules in the air dive back into the liquid. What matters is the net difference between the rate of those jumping out and the rate of those diving in. The warmer the liquid, the faster the rate of evaporation and vice versa.

Flammable vapor is only combustible when mixed with a certain amount of air. The range of the amount of vapor in air that makes a flammable vapor combustible is called the *flammable range*. The flammable range is measured as the percentage of vapor in air. The endpoints of the flammable range are called the *flammable limits*. For example, the flammable range for propane is 2.4–9.6 percent propane in air. Below the lower flammable limit or above the upper flammable limit, a flammable vapor will not burn.

DOI: 10.1201/9781003684213-7

The *flash point* of a flammable liquid is the minimum temperature at which the rate of evaporation is sufficient to create a vapor-air mixture above its lower flammable limit. In other words, the flash point is the minimum temperature at which the vapor–air mixture is ignitable.

Of course, if a source of heat is brought near a cold flammable liquid, the heat will locally increase the liquid temperature and increase the rate of evaporation, so the vapor-air mixture at that point will become ignitable. The local flame will quickly spread heat and increase evaporation so the entire surface will ignite.

To ignite, a flammable vapor-air mixture must either contact a hot object that is above the mixture's *ignition temperature* or the whole mixture must be heated to that temperature. (This temperature is also sometimes called its "auto-ignition temperature" because if a flammable mixture is heated to its ignition temperature, the mixture will simply burst into flame.)

For a fire involving a pure flammable liquid, the only information necessary to analyze the cause of a fire is the liquid's flash point and its ignition temperature. For a fire involving a mixture of flammable liquids, the problem might seem more challenging, but in fact, it is not. It is sufficient to consider each constituent liquid as if it were the only liquid present. A mixture of flammable liquids creates ignitable flammable vapor when any one of the constituents reaches its own flash point.[1]

Here is an example of forensic analysis involving this precept.

Printing Press Fire

A fire in a large printing plant damaged an expensive, eight-color, roll-fed (web-type) press. The press utilized ink whose solvent consisted of 90 percent ethyl alcohol (ethanol) and 10 percent propylene glycol monomethyl ether (glycol ether). Each color was printed on the web at a separate printing station. Immediately after a color was applied, the web was bathed in 230 °F air to dry the ink to prevent smearing at the next printing station. The hot air was produced by a gas-fired furnace and ducted to each station. The air temperature was governed by a thermostat, and the furnace also featured a high-limit thermostat to shut the furnace off if the air temperature reached 300 °F.

A sheetmetal trough was provided below the printing stations to catch ink drippage. Because the ink solvents were flammable, each printing station was equipped with an exhaust system to draw away flammable vapors. However, even though ink droplets would fall and the vapors were heavier than air and thus would tend to sink, no exhaust system inlets were provided at the drip trough itself. Management had apparently decided (incorrectly) that the dripped ink in the trough was not hazardous because the drive motors for each station were encapsulated and sparkproof, the warm air system would never get hot enough to ignite vapors, and no other sources of ignition existed near the press.

Furthermore, the press was equipped with a CO_2 fire extinguishing system that would automatically activate if a fire occurred at the press.

Yet, a fire did occur at the press, which the extinguishing system failed to suppress.

The fire broke out in the drip trough while the operators were filling the printing stations with ink in preparation for a new run. The warm air furnace had been turned on a short while earlier so that drying air would be up to temperature when the press was started. Just before the fire, an operator noticed that the air temperature indicated on the control panel was far higher than normal – 360 °F instead of 230 °F. He also noticed that someone had raised the high-limit setpoint on the second (safety) thermostat from 300 °F to 450 °F, which is why the furnace continued running when the air temperature was so hot. The operator immediately shut off all power to the press and furnace, but fire had begun in the trough seconds earlier.

Forensic investigation found that the press was heavily damaged, but the hot air furnace was not. Testing found that the furnace was so powerful that the air temperature reached 40 °F above the thermostat setpoint (regardless of what the setpoint was) before the thermostat was able to react to shut off the burner. It was also found that the electric valve on the gas line, which the thermostat controlled, was stuck in the open position.

The ignition temperatures of ethanol and glycol ether are 685 °F and 460 °F, respectively. If the ignition temperature were simply a weighted average of the constituents, the ignition temperature would be 662 °F, but this is irrelevant. What matters is whether the vapor-air mixture of *either* of the constituent flammable liquids is within its *own* flammable range when the vapor contacts a source of ignition above its *own* ignition temperature. The vapor density in a vapor-air mixture is measured by the *vapor pressure* of each constituent liquid, which is its propensity to evaporate. The vapor pressure of ethanol at room temperature is 78.4 mm Hg, the air pressure at room temperature is 760 mm Hg, and the concentration of vapor in air is measured as the ratio of its vapor pressure relative to air pressure. Thus, the concentration of ethanol was 78.4/760 = 10.3 percent. The flammable range of ethanol is 3.3–19.0 percent, so the ethanol vapor was ignitable. However, the ignition temperature of ethanol is 685 °F, which is far hotter than any possible ignition source in the area.

The vapor pressure of glycol ether is 7.6 mm Hg at room temperature, so its concentration in air is 7.6/760 = 1 percent. The lower flammable limit of glycol ether is 1.8 percent, so the glycol ether should not be ignitable at room temperature, no matter how hot a nearby ignition source might be. However, glycol ether was in fact what fueled the initial fire. Vapor pressure increases with temperature, doubling about every 18 °F. Thus, glycol ether vapor reaches its 1.8 percent lower flammable limit when the liquid reaches 85 °F. In this case, the hot airstream heated the ink well above 85 °F, so the vapor pressure of the dripped ink was easily above glycol ether's lower flammable limit. The ignition temperature of glycol ether is 460 °F, which the hot air reached a moment before the operator shut off power. The glycol ether ignited, which of course ignited the ethanol. The CO_2 system failed to extinguish the trough fire because the system had no nozzles aimed at the trough.

Oil Burner Malfunctions[2]

Oil-fired equipment is commonly used for home heating. If an oil burner malfunctions, the damage can be costly. The reason is that aside from fire, many kinds of oil burner malfunctions result in soot emission into the premises, and soot is notoriously pervasive and difficult to clean up. The basic reason for the propensity for soot generation is the inherent difficulty in vaporizing fuel oil.

Like any flammable liquid, fuel oil must be vaporized in order to burn. Fuel oil is typically stored at room temperature, so heat might be applied to the oil before being introduced into an appliance's combustion chamber. However, doing so might require a costly mechanism, like an auxiliary heater. Also, oil would have to be kept hot when the appliance was not operating in order for the appliance to start instantly as soon as electric power was turned on. Therefore, instead of pre-heating the oil, the oil is pressurized and sprayed through a fine nozzle into the combustion chamber. The spray breaks up into tiny droplets, which can be ignited by a spark. The heat from the burning droplets quickly vaporizes succeeding droplets. The oil is pressurized by an oil pump attached to the burner, and the spark is created by an ignition transformer that converts 120-volt household current into 10,000-volt current, which is necessary for a spark to jump across the ignition electrodes adjacent to the nozzle. Combustion air is supplied by a fan inside the burner.

Although the burner configuration is simple, in practice, various things can go wrong that prevent smooth ignition and efficient, soot-free combustion. For soot-free combustion, it is essential that the oil vapor be completely intermixed with combustion air. Any portion of the oil spray that is not fully intermixed with air will not burn completely, leaving unburned carbon (soot) in the exhaust. Nozzles come in two types, depending on the form of the cone-shaped oil spray they produce. In one type, called "solid," the cone is filled uniformly with droplets. In the other type, called "hollow," the droplets form a cone-shaped wall with the center of the cone essentially empty. Which type to use depends on the shape of the combustion chamber and the way in which combustion air is intermixed. With both types of nozzles, some way of deliberately creating intermixing is necessary; otherwise, only the outermost droplets of the cone would fully intermix with the incoming airstream. Therefore, to make the entire cone of droplets intermix, the incoming airstream is swirled by passing through a ring of blades before reaching the droplet cone. Problems that interfere with complete intermixing include a dirty or worn nozzle, which can distort the spray pattern and interfere with complete vaporization of the droplets, a damaged or clogged blade ring, a dirty combustion chamber, or insufficient draft. Draft strength affects the volume of air that the fan can force into the combustion chamber and how well exhaust is drawn out. Lingering exhaust can interfere with incoming airflow, which affects intermixing and can also contaminate the incoming air.

Insufficient or improper service is the root of most problems with oil-fired equipment. Oil-fired equipment requires annual preventive maintenance by a professional service contractor. In normal operation, soot accumulates inside boilers and warm air furnaces as well as in flue pipes and chimneys, which must be cleaned out to assure constant proper draft and efficient combustion. Oil normally passes through a filter between the oil tank and the equipment, so the filtration element inside the filter must be replaced. Nozzles become worn and must be replaced. Controls must be tested. Draft must be adjusted using a draft gauge. Combustion efficiency must be tested using instruments.

Notes

1 *Fire Protection Handbook*, 17th ed., National fire Protection Association, 1991.

2 Oil burner operation and malfunctions is a field of expertise unto itself. The present discussion is a general overview. For complete information, see *Domestic & Commercial Oil Burners*, by Charles H. Burkhardt, McGraw-Hill, Inc., 1969.

Arson

8

CAREFUL OBSERVATION IS KEY

Movie and TV detective dramas usually rely on chemistry, toxicology, pathology and other forensic specialties to prove that a fire was an act of arson. Arson is defined as deliberately starting a fire with criminal or fraudulent intent.

Proving arson traditionally requires three factors: *means, motive and opportunity*. Proving *means* often relies on technology, and a forensic engineer can also supply critical proof.

An arsonist may naively assume that a fire will always destroy all incriminating evidence, but this is not true. The aroma of residual gasoline or other accelerant can sometimes linger in the ashes or its presence can be detected with modern laboratory techniques. The absence of any accidental source of ignition would support a conclusion about the means. However, an accelerant need not always be used. Simply applying a match to combustible material would leave no residue. In other words, technology alone cannot always confirm arson.

Documents and interviews can establish *motive*. Who is the insurance beneficiary? Who is in financial difficulty? Who seeks revenge against whom?

Telltale fingerprints, clothing fibers, tire marks, etc., or an analysis of a person's movements can reveal the presence of a perpetrator at the fire scene which establishes *opportunity*.

Forensic investigation can sometimes provide the necessary proof or disproof. Here are some examples.[1]

"Oil Burner Fire"

A fire occurred in the basement of a large, L-shaped house with oil-fired hot water heat. The fire marshal reported the incident as an oil burner fire. A forensic mechanical engineer was hired to determine the nature of the oil

DOI: 10.1201/9781003684213-8

burner malfunction and if an outside party was responsible. The engineer confirmed that a fire had occurred at the oil burner. He photographed the boiler and burner, then disassembled the burner to inspect its components. He found that the fire had clearly been centered inside the oil burner, the cause of which is usually a fuel oil leak at the oil pump's shaft seal. An oil burner uses a motor attached to one side of the body to drive the combustion air fan inside the body and the oil pump attached to the other side of the body. The pump supplies high pressure oil to the nozzle which atomizes oil for combustion. The shaft seal prevents pressurized oil from leaking into the burner body. A shaft seal failure, usually from wear out, allows oil to leak into the burner body where it can be ignited by a spark at the output terminals of the ignition transformer situated on top of the burner body.

The forensic engineer concluded his work and was preparing to leave when the insurance adjuster who had accompanied him asked if the engineer had also seen the other oil burner. "What other oil burner?" "The oil burner around the corner."

It turned out that the house was big enough to be heated by two oil-fired boilers. One boiler was located in each leg of the L-shaped basement, so one was not visible from the other. Both oil burners showed identical fire damage. There was no intervening fire damage to indicate that the fire had spread from one burner to the other, so the only explanation for two simultaneous identical fires is that the fires were deliberately set. At an oil burner, a fire using a liquid accelerant such as kerosene, which is chemically identical to fuel oil, is usually indistinguishable from an oil leak fire within the oil burner.

Gas Leak

The clubhouse of a golf and country club was under renovation one summer when a fire was discovered at a month-old gas-fired warm air furnace and air conditioning unit on the roof. The fire was extinguished before major damage occurred. The fire marshal found the maximum damage was centered at a loose pipe connection (union) on the gas supply line a few inches from the unit.

The gas piping had been pressure-tested when it was installed and had been passed by local building inspector.

Interviews with employees and contractor personnel indicated there had been no need for anyone to have been on the roof at the time of the fire. The renovation contractor was non-union, but the heating subcontractor was a union shop, yet no antipathy or hostility had been reported between union and non-union personnel during the course of the job.

Forensic engineering investigation confirmed that the most intense fire heat was experienced at and near the loose gas pipe connection. Inside the heating unit, mild heat damage was limited to the area adjacent to the loose pipe connection. Nothing abnormal was observed among the unit's components. The loose connection itself had been removed shortly after the fire, so it was not possible to examine the connection for possible defects. Significantly, no accidental source of ignition existed at or near the leaking connection. The furnace was not operative because it was summertime. Therefore, the forensic engineer concluded that the fire was an act of arson.

"Faulty Chimney Pipe Installation"

The owners of a medium-sized, one-story house had a gas log unit installed in the livingroom fireplace. The fireplace was vented via a double-wall metal chimney pipe that extended up through the living room ceiling, the attic and the roof, and terminated with a steel vent cap. A few years later while the fireplace was in use, a fire occurred, which produced maximum damage where the chimney pipe passed through the roof. The attic suffered lesser fire damage and the rest of the house suffered only smoke damage. The fire marshal was unable to find any specific evidence that the chimney pipe had caused the fire, but no other possible source of ignition was apparent. Therefore, even though the chimney pipe was intact from the gas log fireplace to above the roof, he attributed the fire to faulty installation of the chimney pipe, which he believed had allowed hot exhaust to ignite nearby roof structure.

(This is referred to as a "default" conclusion – a deduction without a specific evidentiary basis. Although deduction is a valid analytical principle, for a default conclusion to be valid, it is essential that all other hypotheses are ruled out, which the fire marshal failed to do in this case.)

Several metal components must be provided to assure a chimney pipe is spaced safely away from combustible structure and will remain in place permanently. The chimney pipe must be double-walled so the interwall space provides special separation from the hot inner wall. The interwall space must be vented so hot air from the space convects upward and safely out of the premises. A large, square sheetmetal collar must surround the chimney pipe with its edges fastened to the roof structure to assure the chimney pipe cannot shift or tilt into contact with the structure. In some installations, additional metal braces are also necessary to further support the chimney pipe. Finally, a metal cap must be installed on top of the chimney pipe to prevent wind-induced downdrafts that might obstruct exhaust escape.

Forensic investigation confirmed that the chimney pipe was intact and even though all surrounding roof structure was burned away, all required metal installation components were present. In other words, no evidence existed to explain how the chimney pipe or its installation could have started the fire.

The fire marshal had taken exterior photographs of the house before going inside to inspect the fire scene, one of which revealed the true cause of the fire. At one end of the house, the attic overhung the first floor by several feet to shelter firewood stacked on the ground below. The firewood had been used in the fireplace before the gas log unit was installed. The photograph revealed a narrow, vertical, black band extending from ground level to the top of the stack. The band was obviously caused by a fire that began at the base of the stack and burned up through the overhang into the attic. In the attic, the chimney effect drew the fire toward the chimney pipe, where it ignited the adjacent roof structure. Once the roof was breached, the chimney effect grew stronger, which intensified the fire at the roof pass-through.

The anomalous burn pattern on the firewood stack clearly indicated that the fire had been deliberately set at the base of the firewood stack. In other words, the fire was an act of arson.

Note

1 A third example is found in the chapter on Combustion Venting and the Chimney Effect.

Soot 9

EASILY GENERATED, PERVASIVE AND DIFFICULT TO CLEAN UP

Soot is carbon in the form of tiny particles a fraction of the width of human hair. Soot is generated by the combustion of almost any kind of hydrocarbon.

A hydrocarbon fuel – oil, natural gas, propane, paraffin, wood, plastic, rubber, etc. – consists of hydrogen and carbon. Both will burn when the fuel is ignited. Hydrogen burns readily to H_2O, which is harmless water vapor, while carbon ideally burns to CO_2 (carbon dioxide) and CO (carbon monoxide), which are both noxious gases but usually dissipate readily. However, especially if sufficient oxygen is not available for combustion, combustion will not be complete so that some carbon will not burn at all, creating soot. Soot is an especially pervasive, intrusive and persistent physical contaminant.

Soot's harmful behavior is manifold. Soot particles are so small that they can waft widely from the combustion site and easily infiltrate cracks, crevices and porous surfaces such as fabrics and carpets, even those remote from the combustion site. Soot is electrostatically attracted to plastic. Soot typically does not dissolve in ordinary cleaning liquids, so ordinary washing often will not remove soot.

Perfect, soot-free combustion requires that the necessary amount of oxygen be perfectly mixed with the fuel. In real life, these conditions rarely exist. In accidental fires involving a variety of combustible substances, fuel-air intermixing is usually imperfect, which leaves soot and unburned fuel among the by-products. If the burning substances include oil, paraffin or plastic, soot will often be oily, which makes it all the more difficult to clean up. Soot can be wiped off smooth, non-porous surfaces, but on other surfaces, the only options are to paint over the soot or discard the object. Soot can also coat internal passages of heating system furnaces and boilers, which impedes heat conduction and decreases operating efficiency, thus increasing annual costs.

DOI: 10.1201/9781003684213-9

Fortunately, flame color indicates when combustion is producing soot. Perfect, sootless combustion creates blue flames and imperfect, sooty combustion creates yellow-orange flames. For instance, wood flames are yellow-orange and generate soot. This is apparent on the inner walls of a fireplace, inside the chimney and sometimes on the mantle above the fireplace. Natural gas flames at a stove or heating appliance burner are blue and soot-free, but if the burner flames are flecked with yellow, each yellow fleck produces soot, which over time can accumulate unless the exhaust is vented out of the premises. Gas log fireplaces are deliberately designed to produce yellow flames to simulate wood flames, so gas log fireplaces always produce soot. For this reason, unvented gas log fireplaces will always cause soot damage, while vented gas log fireplaces usually will not.

Candle flames are yellow and therefore generate soot, but are small, so the quantity of soot produced by a single candle is small. However, prolonged use of multiple candles will result in soot damage. How long it takes for soot to produce noticeable damage depends on how well the premises are ventilated. For instance, it has taken centuries for soot to have darkened the interiors of drafty, unheated medieval cathedrals. (Notice the damage the next time you visit Europe.)

Scented candles have an aromatic oil mixed into the paraffin, which tends to burn less completely than paraffin, so scented candles tend to generate more soot than plain candles. Excessive use of scented candles can cause serious soot damage in a house, but even plain candles can cause soot damage over a long period of time.

Portable kerosene heaters generate considerable soot because their flames are wick-fed like candles and completely yellow.

The key to perfect, soot-free combustion is premixing air into the fuel. Air premixing is intended to assure that all fuel is adequately oxygenated to burn completely with blue flames. Insufficient air premixing will cause flames to be less than completely blue and generate soot. For instance, the flame from a stove burner or gas log fireplace will become more yellow if the burner's air intake is obstructed by dust or foreign matter.

Sooting losses are extremely costly because soot is so pervasive and difficult to remediate. If a house is sooted, the entire house may be affected – every surface and object – walls, ceilings, floors, carpets, furniture, cabinet and drawer interiors and contents, closets, appliances, etc. All must be cleaned, repainted or discarded. Otherwise, soot will gradually re-elevate and re-pervade the premises.

The feasibility of subrogation[1] in soot losses depends on the source of the soot – a fire, an appliance malfunction or defective design or service of an appliance. Examples of defectively designed appliances include unvented gas log fireplaces, portable kerosene heaters and scented candles. Liability is also possible if an appliance's combustion air intake is difficult to clean by design or if the maintenance instructions are deficient.

Kitchen Fire Soot

One summer, a fire occurred in the kitchen of a long, narrow, ranch-style house. The kitchen was at one end and the master bedroom was at the other, far away from the kitchen. The fire was caused by an electrical problem in the receptacle behind the refrigerator on the end wall of the kitchen. The fire was sooty because the rubber and plastic insulation on wiring was involved. The fire department arrived quickly and extinguished the fire while it was still behind the refrigerator.

The homeowners found the walls soot-blackened in the kitchen and progressively less blackened with distance from the kitchen. In the master bedroom at the far end of the house, the walls did not seem blackened at all. A small, three-drawer chest in the room was full of winter clothes. The clothes in the top drawer were somewhat sooty – the homeowners' hands became dirty when they handled the clothes. The same was true of the clothes in the middle drawer, but less so. The same was true of the clothes in the bottom drawer, although even less so. When the clothes were removed for laundering or dry cleaning, an unopened package of a dozen golf balls was discovered in the bottom drawer beneath the clothes. The package was cellophane wrapped, and the cellophane was black. The cellophane was removed to reveal three trays of four balls, each wrapped in cellophane, which was black. They opened one of these packages and the balls were black.

Soot is indeed ubiquitous and pernicious.

Scented Candles

A retired couple lived in a modest, two-story suburban house. The heating system and the kitchen appliances were all-electric, and there was no fireplace. They did not usually burn candles, but they occasionally received jarred scented candles as hostess gifts, which by coincidence were all the same brand.

One day, the wife cooked some cabbage for dinner, which produced an unpleasant odor throughout the house. To counteract the odor, she lit three scented candles on the kitchen counter before she and her husband went to bed. The next morning, they discovered the entire house was contaminated with soot. The edges of the candle jars were black.

No warnings were provided for the candles, so the homeowners were unaware that candles, especially scented candles, generate soot. Their insurance company paid for the house restoration and sued the candle manufacturer for the cost.

Gas Log Fireplace

Gas log fireplaces produce soot because they feature yellow-orange flames that resemble wood flames. Yellow-orange flames are obtained by limiting the amount of air mixed into the gas stream inside the burner. Any further air limitation makes the combustion even sootier. Even if an unvented gas log fireplace is used only infrequently, it can generate enough soot to require professional restoration of the premises.

In most gas log fireplaces, the burner consists of one or two horizontal, one-inch-diameter steel tubes with perforations along the top. The tubes are situated below the spaces between logs on a grate above the tubes. Gas is supplied by a small nozzle at the end of a tube. Adjacent to the nozzle are air intake openings, which are adjustable but usually factory-set. Air is drawn into the intakes by the venturi effect created by the velocity of the gas spurting from the nozzle. Over time, dust in the incoming air gradually accumulates on the edges of the air intakes, which gradually narrows the intakes, thereby reducing the amount of air that mixes into the gas. The gas-air mixture becomes richer, making the flames yellower and sootier. Cleaning the air intakes is complicated by the fact that the intakes face rearward, a mirror is necessary to see them, and accessing them is difficult. An ordinary vacuum cleaner hose does a great job of pulling the dust out, but the logs and grate must be removed, and a bit of contortion is needed to position the hose correctly. Thus, the burner design is defective in a way that increases soot damage in the premises, particularly with unvented gas log fireplaces.

Note

1 A type of lawsuit brought by an insurance company to recover the money paid to an insured person or company for damages from an incident. Subrogation is similar to a plaintiff lawsuit.

Steam and Other Boiling Liquid Explosions

10

ENORMOUS POWER, ENORMOUS DANGER

When we think of steam, what comes to mind for most of us is the condensation in our shower stall, the plume from our teakettle, the aroma of chicken soup, or an old-time railroad whistle. These are all harmless, but in fact steam is a powerful force that can be catastrophic.

Steam is water vapor. Water vapor is invisible, like humidity, or visible if the vapor condenses into droplets. All liquids vaporize at a rate that depends on temperature and pressure. Vaporization, also known as evaporation, consists of molecules migrating from a liquid into a vapor. Condensation is the opposite process where molecules transform from vapor into liquid. Both processes occur simultaneously above any liquid. The net effect – loss or gain of liquid – depends on temperature, pressure and type of liquid. In a closed container at any given temperature and pressure, liquid and vapor reach an equilibrium where the same amount of liquid is continuously evaporating and condensing.

Deliberate changes in the temperature or the pressure can be useful. For instance, controlled pressure decrease generates force that powers steam machinery such as steamboats and railroad engines, electric generation turbines and industrial machinery. On the other hand, sudden accidental changes in temperature or pressure can cause explosive energy release with disastrous consequences.

At ordinary sea-level atmospheric pressure of 14.7 psi, water boils at 212 °F. At higher elevations where air pressure is lower, water boils at lower temperatures which can prolong cooking. If the pressure is higher than normal

DOI: 10.1201/9781003684213-10

such as in a boiler, water boils at a higher temperature so hotter steam is produced for useful purposes, like powering a turbine. Similarly, water in a pressure cooker boils at higher temperatures, which speeds cooking – that's the virtue of pressure cookers.

Steam that accidentally contacts a person will produce burn injuries that can be every bit as serious as flame burns. Therefore, accidental steam leaks represent a significant hazard. The greatest hazard is the sudden total release of steam pressure – an explosion – if a pressure vessel ruptures or its lid or access panel blows off.

Properties of steam at various temperatures and pressures, known as Steam Tables, are used to quantify the hazardous capabilities of steam.[1] The following examples illustrate the steam explosion hazard.

Pressure Cooker Explosion

The owner was using a six-quart, low-pressure (1 psi) pressure cooker to make soup when the lid suddenly blew off. Steam and scalding-hot liquid erupted onto her and caused substantial burns. She had purchased the pot a few months earlier, had read the directions, and had used the pot of few times prior to the incident.

The pot consisted of a round cast-iron body with a glass, gasketed lid. The lid featured a pressure-limiting valve and a small rubber safety plug calibrated to blow out in case the pressure rises too high. The glass lid had a beveled metal rim. The lid was designed to be secured to the body of the pot by a pair of plastic handles on the body. Each handle had two parts, one attached to the body and one that slides over the lid rim. Although the handles were advertised as "locking," in fact the handles did not lock at all, but rather were free to slide in and out unless the sliders were fully extended onto the lid rim. If less than fully extended, the sliders would rest on the beveled (sloped) portion of the rim, where a portion of the upward force created by pressure in the pot translated into outward force on the sliders. Moreover, even at room temperature and pressure, extending the sliders onto the lid rim was difficult, requiring more effort than a user would expect it to take.

At the time of the incident, the lid was in place on the pot and the pot was half full of vegetables and stock over medium heat. The liquid was boiling and the lid's pressure-limiting valve was moving up and down properly to keep the pressure at 1 psi. Suddenly, the lid came free and the owner and was struck by steam and scalding liquid.

Inspection found all components intact. It was concluded that the lid came free because one or both handle sliders had not been fully extended and had been forced backward by the pressure in the pot. If the pressure was only 1 psi, why was the incident so explosive?

At 1 psi pressure, the liquid temperature was 215 °F. If the pressure suddenly dropped to atmospheric pressure, the liquid would be 3 °F higher than its 212 °F boiling temperature at atmospheric pressure, so the liquid would instantly boil in order to reduce the temperature to 212 °F. Unlike when boiling takes place at the bottom when a pot is heated over a burner, when pressure is suddenly released from a pot at 1 psi, instantaneous boiling occurs *throughout the liquid.* In this case, the sudden steam formation instantly expanded the liquid volume by more than six times, propelling the liquid out of the pot.

The investigation concluded that the pot was defectively designed in many ways, including handles that do not lock, handles that are difficult to extend, and faulty instructions that do not describe how to make sure the handles are fully extended and do not warn about the steam explosion hazard.

Steam Rocket

The power of steam can be illustrated by a hypothetical steam rocket. Take a 6-foot length of 6-inch diameter steel pipe, cap one end and install a nozzle on the other. The nozzle would consist of a plate with a 1-inch diameter hole in the center. Fill the pipe with water and install a tapered plug in the nozzle. The water-filled pipe will weigh about 184 lb. Now, place the pipe nozzle-end downward into a slightly larger diameter, 6-foot-long, vertical steel pipe equipped with a latching lid. Build a fire around the outside so the water reaches 700 °F. The water cannot boil because the slightest boiling increases the pressure, which prevents further boiling. At 700 °F, the pressure will be 3,200 psi. Release the latch and the pressure will instantly eject the nozzle plug and steam will erupt throughout the water. Steam will jet from the nozzle at such high velocity that the pipe "rocket" will develop 2,400 lb of thrust that will propel the pipe to an altitude of 30,000 feet.

Aluminum Smelter Explosion

A 10′ × 8′ × 7′ gas-fired, 1,400 °F furnace was used to melt scrap aluminum in a baseball bat manufacturing plant. The scrap was stored outside in open-topped dump carts called skiffs. The skiffs did not have drain holes. Normally, when brought into the plant, a skiff was placed near the furnace for pre-heating to evaporate rainwater before it was dumped by forklift into the furnace. On one occasion, however, a skiff that had been outside for ten days was brought into the plant and immediately dumped into the furnace. Over the preceding ten days, the area received an inch of rain. When the skiff was dumped into the furnace, the furnace instantaneously exploded, which blew out the furnace walls, blew the skiff 105 feet, blew the forklift 15 feet, killed the forklift operator and another nearby worker, lifted the roof of the plant and damaged pipes and electrical cables.

Analysis showed that an inch of rain would amount to 5.6 gallons of water in the skiff. When introduced into the furnace, the 1,400 °F temperature instantly transformed the water into almost 3,500 cu.ft. of steam. The increased pressure would have been sufficient to destroy the furnace, but the furnace was vented, which attenuated the pressure increase. However, the temperature was so high that the water dissociated into hydrogen and oxygen. The oxygen contacted molten aluminum, which oxidized in a reaction that generated a tremendous amount of heat and the hydrogen burned in air, which generated even more heat. The thermal expansion of the air from the combustion reactions greatly exacerbated the steam expansion, thus producing the incident's tremendous power and catastrophic damage.

Note

1 For example, see G. J. Van Wylen, *Thermodynamics*, John Wylie & Sons, 1960; or R. H. Perry, *Engineering Manual*, 3rd ed., McGraw-Hill, 1976.

Pipe Bursts From Freezing 11

PIPES BURST BECAUSE THEY FREEZE, BUT NOT *WHERE* THEY FREEZE

It happens to household pipes, to factory sprinkler systems, in buildings of all kinds. Water pipes exposed to freezing temperatures can burst.

Copper pipes are more vulnerable than steel pipes because copper pipes are not as strong as steel pipes, but steel pipes can rupture at joints.

Water expands ten percent by volume when it freezes. In liquid form, water molecules are amorphous, meaning disordered and mobile. In frozen form, water molecules are crystalline, meaning ordered and rigid. The transformation from disorder to order causes expansion, meaning the density is lower, which is why ice floats on water.

If the water is unconfined like in a pond or open bowl, ice is free to expand as it forms, but if water is confined in a closed vessel like a tank or pipe, the ice will expand against the vessel. Whether a vessel ruptures depends on the strength of the vessel and whether it contains any air. If a tank or pipe is completely full of water, the expansion will rupture the tank or pipe because the expansion pressure can theoretically exceed 40,000 psi. No ordinary vessel can withstand such extreme pressure.

In practice, the pressure will not get that high because a pipe will "give" a little, a tank or pipe will contain some air, and the ice expansion will compress the air, which reduces the maximum pressure. Still, when a full vessel freezes, the pressure can reach thousands of psi; whether the vessel ruptures depends on the volume of air compared with the volume of water. The less air, the less the ice pressure will be attenuated, and vice versa.

Now consider a water-filled pipe with no air in it at all. If both ends of the pipe are closed off such as by valves, the ice pressure will be immense and the pipe will rupture. For example, the rupture strength of ordinary Type L copper pipes as found in domestic water systems ranges from 7,200 psi for 1″ size pipes to 9,840 psi for 1/2″ size pipes. These failure pressures are far

DOI: 10.1201/9781003684213-11

lower than the expansion pressure when water turns to ice, so the pipes will absolutely rupture.

However, in real life, the situation is usually more complicated. First, an entire pipe may not freeze. Second, air may be trapped in a pipe. Third, freezing entails loss of heat, which takes time and depends in part on how much colder the exposure temperature is compared with the original pipe temperature. Thus, not all pipes exposed to subfreezing temperatures may rupture.

A common situation is where a limited length of a water pipe is exposed to freezing, such as for instance, the last foot or so of a pipe leading from a basement to an outside hose bibb. When the outside temperature falls below freezing (and the bibb is not in direct sunlight), the bibb will certainly freeze, but whether the bibb and pipe rupture depends on whether the ice that forms is able to expand.

Inside the house, only the final foot or less of the pipe will probably freeze, with the rest of the pipe remaining liquid because the house temperature is above freezing. In most domestic water systems, the system is open to the source (the municipal supply), so if water freezes anywhere in the system, the expansion will simply push some liquid water back into the source. A rupture hazard does exist if the bibb pipe has a closed shutoff valve inside the house near the exterior wall. In this situation, the water between the bibb and the closed valve is trapped, so if the bibb freezes, the ice expansion will pressurize the pipe, which can rupture. When the outside temperature grows warmer, water will leak from the rupture, but the closed valve assures that the volume of the pipe is so small that the leak will likely cause scant damage.

Major damage can occur when a pipe ruptures at a point that is fed by a continuous water supply. For instance, suppose a water pipe is exposed to subfreezing air from a nearby drafty window or the pipe is insufficiently insulated, so the pipe freezes in this location only. The ice will form a plug in the pipe. As freezing continues, the plug will grow, pushing liquid water ahead of it. If the end of the pipe is closed off by a valve or another ice plug, the growing plug(s) will create pressure in the trapped water, which can rupture the pipe. In other words, the rupture will occur *somewhere in the section containing trapped liquid water, not where the pipe freezes.* When the upstream ice plug thaws, water can leak continuously from the rupture and cause substantial damage. Thereafter, when the problem is traced, no source of cold air may be found at the leak site, thereby sometimes leading to the incorrect conclusion that the leak was caused by a defect in the pipe. Unless evidence of corrosion is found at the leak site, and especially if the edges of the leak are bulged outward, freezing is the correct conclusion – but not at the burst point. The source of cold air must be sought that caused an ice plug or plugs to form, which created the intense pressure that ruptured the pipe.

Pipes in insulated buildings sometimes rupture from freezing due to a heating system failure or simply insufficient insulation. Insulation does not prevent heat loss; *it only slows the flow of heat.* The capability of insulation depends on the type and thickness. Most materials slow the flow of heat to some degree. Those that do so well are known as insulators. The thicker the insulation, the slower the flow of heat. The greater the difference in temperature between the warm side and the cold side, the faster the flow of heat. Heat flow is faster if the air on the cold side is moving, like from the wind. The insulating capability of insulation is called its "R-factor." The R-factor required to protect a building from freezing can be calculated and the result then used to specify the type and thickness of insulation required.

Vertical Tank Failures 12

VENTING IS CRITICAL

Municipal water systems and rural farms commonly use large, vertical tanks for water storage. Because of the huge quantity of water such tanks can hold, collapse or rupture can be catastrophic. Such tanks are usually filled and drained via pipes at the bottom and have air vents at the top. Air escapes through the vent when the tank is filled and flows in through the vent when water is withdrawn. The following discussion illustrates the critical importance of free air movement through the vent.

Venting is essential for a tank's structural integrity. Assuming that some air exists at the top of a tank, if the vent is clogged, stuck closed or is otherwise obstructed, draining water from the tank will initially be possible because the trapped air will expand to replace the withdrawn water. Expansion means the air pressure will be reduced. Reducing the air pressure below atmospheric ambient pressure – drawing a vacuum – will impose a load on the tank roof. A tank roof is designed to withstand a certain load such as from snow and ice, but of course there are limits. The maximum load a roof can support depends directly on the negative air pressure in the tank. In fact, the critical negative pressure in a tank is surprisingly modest, as the following example illustrates.

Consider a large water tank that measures 20 ft diameter by 30 ft high. The cross-sectional area is 314 sq.ft. If the tank is completely full of water, the pressure at the bottom is 13 psi. Thus, if the tank were unvented, external air pressure – ordinary ambient pressure, 14.7 psi – would be greater than the internal water pressure so outflow through a small pipe would be impossible.

Now, suppose the tank were less than completely full so there was a certain amount of air above the water. If the vent is obstructed, withdrawing water is possible because the air expands as water is withdrawn. As the air expands, its pressure grows lower. The lower the pressure, the greater the external pressure on the roof. For instance, suppose the air pressure was reduced by 1 psi (from 14.7 psi to 13.7 psi). This would impose a load of

DOI: 10.1201/9781003684213-12

45,200 lb on the roof – 22–1/2 tons. For perspective, that much load on a roof would be the equivalent of 4½ to 9 feet of snow.

The volumetric expansion of the air to create a pressure differential is measured as a percentage of the original volume of air. An expansion of only 7.3 percent is sufficient to create a 1 psi differential. Suppose the 20 ft diameter tank contained one foot of air. One foot of air is 2,412 gallons. Reducing the air pressure by 1 psi would require withdrawing .073 × 2,412 gal, or 176 gallons of water. A reasonable, very approximate flow rate for a 4″ size pipe would be 240 gallons per minute. Thus, draining 176 gallons from a 20 ft diameter water tank would take only 44 seconds. If no one were aware that the tank's vent was obstructed, the first notice that the vent was obstructed would be the roof collapse.

The moral of the story is that preventive measures to prevent vent obstruction are critically important for large water tanks. Vents must be screened to prevent insect, bird or animal entry, so vent screens can be obstructed by debris or freezing of rain, sleet and snow. Preventive measures might include a means of heating a vent, or auxiliary vents triggered electrically or mechanically by a slight reduction in tank air pressure. Alternatively, a vent could be piped to a heated building. At the very least, a tank should be equipped with a remotely-monitored air pressure alarm to automatically close the water outlet valve.

Residential Oil Tank Failures 13

STRESS CONCENTRATION TEARS WELD SEAMS

In a home with oil-fired heat, fuel oil is commonly stored in a steel tank, usually in the basement but sometimes outside. Ordinary steel tanks are 275 gallon-size made of welded steel. These tanks are racetrack shape ("elliptical") when viewed from the end, with straight middle portions of the sides and semi-cylindrical at the top and bottom. The shape is critical with regard to the failure of the tank.

Steel tanks can fail from corrosion over a long period of time. Exterior tanks can corrode from weather and interior tanks can corrode from long-term bacteriological action in accumulated sediment. Bacteria enter the tanks in air drawn in as oil is drawn out. Corrosion will eventually lead to leakage, usually starting with a trace, but corrosion can become so severe that a major hole can break away when the tank is filled and a greater weight of oil is imposed on the bottom of the tank.

A sudden major corrosion leak can immediately empty a tank into the basement, which is obviously catastrophic. Oil is difficult to clean up because it seeps into every tiny crack, crevice and porous object, including concrete. Following gross mop-up, plaster, sheetrock, wooden framing and furniture must be removed and discarded and special detergents must be used to remove oil from concrete. Otherwise, the aroma of oil will persist and make the house unlivable.

OVERPRESSURIZATION

Another type of sudden, catastrophic tank failure is rupture, also called bursting. Bursting is caused by excessive internal pressure, which can affect

DOI: 10.1201/9781003684213-13

even tanks in perfect condition. Excessive pressure usually results from improper filling. Elliptical tanks are much weaker than spherical or cylindrical tanks when overpressurized. In spherical and cylindrical tanks, pressure is imposed equally on all surfaces, like in a balloon. In an elliptical tank, pressure causes the walls to bow outward, but the stress is not distributed uniformly. The racetrack-shaped end walls of the tank include straight seams and curved seams. When pressure inside the tank is imposed on the seams, stress is concentrated where the straight seams transition to curved seams. As pressure increases, a weld at the transition will tear apart, releasing as much as 200 gallons into the premises.

Overpressurization is caused by a venting problem when a tank is filled. Ordinary steel fuel oil tanks are not designed to withstand significant pressure. Such tanks are tested to 5–7 psi at manufacture, and if the tank were completely full – which it normally never is – the maximum pressure after filling would be only about 1.35 psi at the bottom from the weight of oil.

When a tank is filled, the incoming oil displaces air. Therefore, a tank must be vented to allow the air to escape. A residential tank is filled through a pipe from outside the house, and a separate vent pipe carries air from the tank out of the house. By contrast, an automobile gasoline tank has only a single pipe to carry incoming gasoline and outgoing air, but this arrangement is not safe for residential fuel oil tanks. A gasoline hose nozzle fits loosely into the fill pipe, which allows air to escape around the nozzle if necessary. In residential oil tanks, the hose nozzle locks onto the fill pipe outside the house so air cannot escape, hence the need for a separate vent pipe. The locked-on nozzle prevents oil from leaking, which would contaminate the ground below the fill pipe. By contrast, in a gas station, spilled gasoline would fall onto concrete and quickly disappear because gasoline evaporates quickly. Fuel oil evaporates only slowly.

The hose nozzle at a gas station has a safety feature that automatically shuts off the nozzle when the tank is full. Fuel oil delivery nozzles do not have that feature. Instead, to prevent overfilling the tank, a tank has a device in its vent pipe that generates a whistle sound when air passes through. The whistle is designed so that the sound automatically ceases before the tank is completely full, so the cessation tells the deliveryman to shut off the nozzle. The whistler, also known as a vent alarm, consists of a slender, 6″ long tube that sits inside the vent pipe where the vent pipe joins the top of the tank. The tube works like a penny whistle. The tube is of a much smaller diameter than the vent pipe, so escaping air passes both through and around the whistle tube. When the rising level of oil reaches 6″ from the top of the tank, the oil covers the bottom of the tube, so the whistle cuts off. Meanwhile, air can still escape from the tank around the outside of the tube. When the whistle halts, the deliveryman knows to shut off the nozzle, at which point the oil level is about 25 gallons below completely full. The flow rate of the average delivery hose is 20–50 gallons per minute, so the deliveryman has at least half a minute to shut off the nozzle. Proper delivery procedures require the deliveryman to remain with the nozzle during the entire delivery so he can shut off the nozzle as soon as the whistle stops. Thus, in principle, the whistler should assure that a delivery halts well before a tank becomes totally full.

There is no gauge outside a house to tell a deliveryman how much oil a tank contains. He cannot know whether it will take ten minutes or half a minute to fill the tank. Thus, it is hazardous for him to leave the nozzle even for a moment. Of course, mistakes happen. A deliveryman might leave the nozzle for some reason. If inflow continues despite the absence of a whistle, the result can be catastrophic. If the vent pipe is clear, the least problematic result is that a small amount of oil will be discharged from the vent pipe when the tank becomes full. The ground will be contaminated, but the quantity involved will be small as long as the deliveryman is present to immediately shut off the nozzle. If he is not, the resulting exterior contamination could quickly become severe.

OIL TANK RUPTURE MECHANISMS

Alternatively, the tank can rupture and release a huge quantity of oil into the house. There are two types of failure mechanisms for a tank rupture, either of which can occur if inflow continues without a whistle. One is simple overpressure, described earlier, which will rupture a weld seam if the vent is obstructed. The tank will simply "inflate" to the point of failure.

The failure mechanism is hydraulic shock. Hydraulic shock, also known as water hammer, occurs when fluid motion is suddenly halted or changes direction. Think of the surface area of the oil level in a tank as the level rises during a delivery. When the tank is less than about three-quarters full, the surface area is the same as the overall length and width of the tank (its cross-sectional dimensions at mid-height). When the oil level reaches the height where the upper, semi-cylindrical portion of the tank begins to curve over, the width of the tank grows progressively narrower, so the surface area grows smaller.

Now think of the air above the oil. As the oil level rises into the semi-cylindrical part of the tank, the volume of air gradually diminishes, and because the inflow rate is constant, the oil level rises faster and faster. Eventually, the air becomes just a bubble centered on the vent outlet. As the oil level continues to rise, the bubble shrinks and the oil converges at high speed on the vent outlet. When the converging oil reaches the vent outlet, meaning when all air has escaped, the oil flow suddenly changes direction from horizontal convergence to upflow into the vent pipe. The sudden convergence creates a shock in the oil. The shock, which is a sharp pressure pulse, can have the same effect as slow overpressurization – the shock will impact the walls of the tank and can be forceful enough to rupture an end seam.

Thus, oil tanks can rupture whether or not the vent is obstructed. The key factor in both situations is inflow without a whistle. It is therefore axiomatic that *NO WHISTLE, NO DELIVERY.* It follows that unless a tank ruptures from corrosion, the cause is defective delivery.

Control Standards 14

POTENTIALLY HAZARDOUS IF OUTDATED

In designing combustion equipment of any kind, whether a large steam boiler or an ordinary kitchen range, the key axiom for control logic is *ignition before fuel*. That is, the source of ignition must exist before fuel is introduced. The purpose of this rule is to prevent the possibility that fuel accumulates before it ignites. An explosion is the ignition of a mass of unburned fuel. In other words, to prevent an explosion, proper design requires that fuel ignite immediately as soon as it is introduced.

PUFFBACK HAZARD

When this precept is violated – when fuel accumulates and then ignites, the result is called a delayed ignition explosion. Delay ignition explosions of small amounts of fuel can be minor. For instance, delayed ignition explosions in oil-fired home heating appliances are manifest as audible bangs called "puffbacks." Such puffbacks are of limited magnitude because the combustion chambers are small. The most common consequence of a puffback is a burst of soot from the heating appliance. Soot can be widespread and difficult to remediate, but puffbacks usually do not damage the equipment. Puffbacks are usually caused by spark generation problems such as a weak ignition transformer or short-circuiting by soot on the transformer terminals.

A delayed ignition explosion in a gas-fired domestic kitchen appliance manifests as a burst of flame. While not injurious to the appliance, a burst of flame can ignite nearby combustibles such as user's clothing.

Delayed ignition explosions in larger equipment such as industrial steam boilers can be far more powerful and capable of damaging the equipment

DOI: 10.1201/9781003684213-14

and its surroundings as well as injuring nearby personnel. Large types of equipment have elaborate safety devices built into their operating controls. The safety devices typically use an optical flame sensor (light detector) to detect the presence of flame in the combustion chamber. When the burner starts, the controls require that flame (main flame or pilot flame) be established within a certain time interval, or else the burner shuts off. Similarly, after flame is established, the burner will shut off if the flame disappears for any reason.

The quicker the optical flame sensor signals the principal operating controller, the smaller a delayed ignition explosion can be. Modern optical flame sensors are extremely sensitive and can react almost instantaneously to flame presence or absence. How fast an operating controller then reacts is called the Flame Failure Response Time (FFRT). Early in the history of operating controllers, the FFRT was very slow, about 30 seconds. Second-generation controllers provided 4-second FFRTs, which is a lot faster but still not as fast as needed. The latest generation of controllers provides FFRTs of less than one second, which is fast enough for safety.

STANDARDS

Nationally recognized standards exist that govern the design of large combustion equipment to minimize the possibility of delayed ignition explosions. (The same principle – ignition before fuel – also applies to small domestic appliances, even though design standards that explicitly codify this principle may not always exist.) For large boilers, the applicable standard is formulated and promulgated by the American Society of Mechanical Engineers (ASME) as Publication CSD-1, "Controls & Safety Devices for Automatically Fired Boilers." The standard is accepted as an American National Standard by the American National Standards Institute (ANSI). The 1977 and 1995 editions of CSD-1 both required an FFRT of *not more than four seconds*. Yet, by well before 1995, controls were on the market that provided a 0.8-second FFRT.[1]

Obviously, the shorter the FFRT, the less unburned fuel can accumulate, so the less powerful an explosion will be if the fuel ignites. With a 0.8-second FFRT, an explosion could be only one-fifth as powerful as with a 4.0-second FFRT, so the likelihood of operator injury or equipment damage would be greatly reduced.

Boiler Explosion

The problem is that national safety standards may not always be kept up to date. A perfect example concerned a boiler explosion in a large, commercial vehicle repair facility. Heat was provided by a pair of large, 7 million BTUH (British Thermal Unit), low-pressure, gas-fired steam boilers, both manufactured in 1980. Both were equipped with safety controllers that comported with CSD–1. Both boilers were well maintained, were regularly inspected for state licensing and were operated by trained, experienced, full-time technicians. The boilers were more than 20 years old when one day, a valve stuck open on the gas supply line to one of the boilers, which resulted in a powerful explosion that blew the boiler apart and caused steam burns to the operators. Investigation confirmed that the incident consisted of a combustion explosion, not a steam explosion such as might be caused by excessive pressure or a thermal shock of some kind.

The incident occurred when the boiler was automatically restarting after having been properly shut off due to low water level, when the operators drained water to purge accumulated sediment. The purging was a normal, proper safety procedure that was performed every day and had never resulted in a problem. The incident was a delayed ignition explosion which in principle should never have been large enough to cause damage.

Investigation revealed that the controller's FFRT was 4 seconds rather than 0.8 seconds, which had been state of the art for many years. The controller was old and had never been upgraded. Upgrading could have been mandated for state licensing, which requires conformance with governing standards like ASME CSD-1, but CSD-1 still allowed a 4-second FFRT – which ASME in fact confirmed. How was this possible? The chairman of the relevant ASME committee admitted that the need to upgrade CSD-1 to a far faster FFRT existed for many years before the incident, but the need to revise the standard had simply "fallen through the crack."

The moral of the story is that *compliance with safety standards does not always assure safety.* Particularly for large, powerful, potentially dangerous equipment, safety assurance requires analysis of all operating parameters, equipment and procedures in each specific application. A safety standard should be treated as a guide, not as a guarantee of safety.

Note

1 For example, Honeywell Type R4140L1105 Burner Control Module.

Conclusions, Opinions and Liability

15

THE BOTTOM LINE

After a forensic mechanical engineer has completed his investigative work, the results usually must be set forth in writing. A report should begin with the actions the engineer took in the investigation, interviews conducted, documents reviewed, etc. The next section should describe the evidence and information obtained, including photographs.

Next, the report should "reconstruct" the incident – describe in detail what actually happened according to the facts and information. The analysis should be described in detail, including calculations and logical deductions. Sometimes, especially when a machine did not malfunction, the reconstruction may be simple and obvious. For instance, a machine operator might have been distracted so that he inadvertently reached into a punchpress just after stepping on the actuation treadle so he lost a finger when the ram came down. In another case, the operator may have been knocked unconscious and was unable to recall exactly what happened. It is then up to the forensic engineer to piece together the accident sequence based on known facts and engineering and scientific principles.

The next-to-the-last section should summarize the conclusions the investigation reached about causal defects.

ANALYSIS

When all the facts and information are assembled, a rigorous procedure must be used to reach a conclusion about what happened and why. The

DOI: 10.1201/9781003684213-15

procedure is known as the Scientific Method. The Scientific Method entails postulating hypotheses that explain the incident and then evaluating how well each hypothesis matches the facts and scientific laws and principles. Hypotheses that do not match well are cast out, ultimately leaving the one that matches best.

It is important to recognize that the hypothesis that matches best still may not match all the facts exactly. Best does not necessarily mean perfect. All facts may not always be completely known, yet they may be the best information available. For instance, important evidence may be consumed in a fire or the person involved in an accident may have perished. Sometimes the remaining hypothesis may require arcane reasoning or physical principles that are not well known to those without technical education, such as attorneys and insurance people. In such circumstances, it is well to keep Ockham's Razorl in mind. Ockham's Razor that states that when no other means exists to decide between two equally likely hypotheses, the simpler one is usually correct. Although not always true, it is more often true than not.

RESPONSIBILITY VERSUS LIABILITY

A given incident may involve many parties, each of whom may or may not be responsible. For instance, parties to a machine accident might include the injured person, the worker's supervisor, the employer, the machine manufacturer, the company that sold the machine, a service contractor who repaired the machine, the company that owned the machine (if not the employer), the medical facility where the worker was treated, and the rehabilitation facility where the worker went for therapy. Not all of these parties might play a role in a given accident, and it is the forensic engineer's job to sort out which parties did and did not. This function is referred to as assigning *responsibility*.

However, responsibility is not the same as *liability*. Liability means legal responsibility, which is usually the principal concern of most parties following an incident. The difference between responsibility and liability is that some responsible parties may be immune from liability. For instance, employers are usually required to carry workers compensation insurance. Such insurance pays injured workers and/or pays for medical care. In exchange for carrying workers compensation insurance, an employer is legally immune from liability for the accident. The reasoning is that an employer would almost certainly be involved in an accident in some way, such as by inadequate training, supervision, maintenance, or repair of the machine, and poor lighting or environmental conditions. Were it not for what's called this "comp bar," an employer might be bankrupted by an accident which would harm not only the employer but also stockholders, banks, suppliers and other employees.

However, the "comp bar" has limitations. An employer might not be immune from liability if he was the designer of the defective machine involved in an accident, not just the employer; or if the employer deliberately defeated a safety feature that was an original feature of the machine.

Another responsible party that would be immune from liability would be a party that is absolved from liability by contract. For instance, a service contractor might be contractually immune if his contract includes a hold-harmless provision.

ASSUMPTION OF RISK

A machine manufacturer, seller or service contractor would not be considered liable if the action taken by the injured party was extraordinarily dangerous. Extraordinarily dangerous means the action was beyond the range of actions that might be reasonably expected during proper operation of the machine, even considering ordinary variations in procedure. Taking an extraordinary action would be termed *assumption of risk.*

Examples include horseplay, being under the influence of alcohol or drugs, or using the machine for a purpose for which it was not intended, like using a blowtorch to light a cigarette. Also qualifying as assumption of risk would be deliberately removing a guard over the danger zone where the accident occurred, or a maintenance worker deliberately entering a danger zone without taking precautions to assure the machine was inert. A maintenance worker is assumed to have been better trained about safety than an ordinary operating worker and thus is held to a higher standard.

Assumption of risk also applies if an accident occurs because the user failed to read or heed operating instructions. This assumes the instructions were complete and comprehensible, and the user had access to them, which is not always true.

Hydraulic Press Accident

A worker in a ball bearing manufacturing plant incurred a severe hand injury when the ram of a large hydraulic press descended while his hand was inside the danger zone. The press was used to bend thick steel bars into circles, which would later be welded and machined into ball bearing races. The press was extremely powerful but slow-acting. Ram's movement up and down was controlled by a valve on the hydraulic pressure pipe. The valve was manually operated using a tall, vertical lever located close to the right side of the press. Pushing the lever caused the ram to descend; pulling the lever caused the ram to rise.

The operator sits on a chair in front of the press and takes a bar from a nearby stack and places it on a vee-shaped anvil on the press bed. He then pushes the lever forward and the ram slowly descends, forcing a die attached to the ram into the bar, which bends. The operator raises the ram, repositions the bar and brings the ram down again. The cycle is repeated until the desired shape is obtained.

Because the ram moves so slowly, no guard is provided in front of the bed opening, nor is a guard provided to prevent accidental actuation of the operating lever.

One day, an experienced employee was operating the press when he leaned far over to his right in order to talk to a worker on a machine rearward of the press. In leaning over, the employee inadvertently moved the operating lever forward, so the ram descended. He was still holding a bar on the anvil and failed to remove his hand before the ram crushed his hand.

Because the hazard posed by the ram was fully visible and the ram descended so slowly, the hazard was easily avoided. Therefore, it was judged that the accident was attributable to the employee's having assumed the risk of injury by carelessness that went beyond reasonable propensity for error.

Electric Pole-Type Chain Saw Accident

A homeowner purchased a pole-type electric chain saw to trim small branches from a tree in his backyard. The chain saw was cord-powered, 8 inches long, and mounted on an extensible pole with a maximum length of 8 feet. The trigger was located in a housing at the base of the pole. The homeowner had never used such a chain saw before, and this was the first time he ever used the subject saw.

The saw came with complete operating instructions, but the homeowner thought it unnecessary to read them. One of the instructions warned not to use the saw while standing on a ladder – the saw should only be used while standing on the ground.

The tree was tall, so the homeowner took his 24-foot extension ladder, extended it to maximum length and placed it up against the tree. The foot of the ladder rested on the smooth, paved surface of his driveway. He extended the saw's pole to its maximum length, climbed a short distance up the ladder and trimmed some low branches. He then continued climbing and trimming until he was standing on the upper portion of the ladder, approximately 15 feet above the ground. To reach a high, distant branch, he twisted to the right and reached the saw outward and upward. He began cutting the underside of the branch using the top part of the chain saw when a kickback occurred which drove the saw axially toward him and propelled him off the ladder. Impact with the ground resulted in multiple pelvic fractures.

Forensic engineering investigation revealed that the saw was marketed as having an anti-kickback chain, but in fact the type of chain on the saw was actually only a semi-anti-kickback type. The homeowner wished to hold the saw manufacturer, importer and seller liable on the basis that the saw was defective without having a maximally anti-kickback chain.

The homeowner's theory of liability was tantamount to asserting that the accident would not have occurred if the saw had been equipped with a fully anti-kickback chain. However, forensic engineering analysis indicated that this assertion is false. The chain of a chainsaw runs in a loop around the bar, which is a fixed track for the chain. A motor at the base of the bar propels the chain outward on top of the bar, downward around the round tip of the bar and back along the bottom of the bar. A chainsaw kickback occurs when something suddenly halts the chain, like sudden contact with a hidden nail or if the chain is pinched within the cut. The motion of the saw in response to a kickback depends on where in the chain's loop the kickback occurs.

The chain and motor are moving and therefore have inertia. Sudden stoppage of the chain transfers the inertia to the saw structure, propelling the saw in the opposite direction from the direction the chain was moving. If the chain is cutting with the bottom, straight portion of the chain, in other words, during a downward cut on top of a log or branch, a kickback will cause the saw to jump forward, away from the operator. If the chain is cutting with its outer end where the chain turns downward around the tip of the bar, a kickback will propel the saw upward, potentially rotating back to strike the operator. If the chain is cutting with its top, straight portion, in other words, upward into the underside of a log or branch, the kickback will cause the saw to jump backward toward the operator. Thus, the homeowner was cutting upward at the time of the accident.

The theory of liability asserted by the homeowner was incorrect for two reasons. First, most kickbacks occur, and are most dangerous, at the tip of the bar. The greater chance of a kickback at the tip is due to the fact that the space between a chain's cutting teeth increases when the chain passes around the tip. For this reason, anti-kickback chains are designed to be most effective during tip cuts. But in the present case, the kickback did not occur at the tip because if it had, the saw would have rotated upward, not back toward the homeowner. Thus, it was unlikely that a maximally anti-kickback chain would have made any difference in the propensity for the subject kickback.

In fact, there are many reasons why a kickback can occur. Anything that suddenly halts the chain will result in a kickback, including not only unexpected encounters with a harder-than-usual type of wood or a hidden nail, but also sagging of the workpiece that clamps the chain in the cut, or twisting of the saw so the chain binds in the cut. Of these, sagging of the branch while it was being cut up from the bottom and/or twisting of the saw is what caused the kickback in the present case. Recall that the homeowner was standing on the upper portion of a long ladder reaching outward and upward with the saw to cut a high, remote branch. His position was precarious. Is it any wonder that he inadvertently twisted the saw so it kicked back?

The bottom line is that the homeowner *assumed the risk* of injury by failing to heed the instruction provided with the saw, which said that the saw should not be used while standing on a ladder, but rather only while standing on the ground. The homeowner was therefore counseled that there really was no basis for action against the saw manufacturer or any other party.

OPINION

The final section of a report contains its ultimate bottom line, which is the engineer's professional opinion about the incident. The substance of the opinion and the form in which the opinion is expressed are both essential to a report's credibility.

The key content of the opinion is the identification of the party or parties that are liable.

Developing an Opinion

To be legally viable, an opinion must be derived using the Scientific Method, as discussed earlier. Unsupported opinions, called "net" opinions, or opinions not derived by using the Scientific Method, are not valid.

Expressing an Opinion

The opinion must be held and expressed "to a reasonable degree of professional certainty" and stated in those words. "Reasonable degree of professional certainty" means definite or probable. "Possible" is insufficient because it can be argued that anything is possible, so an opinion expressed as possible cannot serve as the basis for legal action.

For legal exactitude, the opinion should be stated as, "The incident would not have occurred if the [device or service] had not been defectively [designed, manufactured, serviced, etc.] by . . . such that" This phraseology places the party directly in the chain of causation regardless of the actions of any other party.

For a report to be most useful, the "Opinion" section needs only to include opinions about which parties are liable. Mentioning parties who are not liable, whether responsible or not, is not usually necessary. An exception might be if the forensic engineer has been retained by the defense, in which case, if the defending party is not responsible, that party should be specifically absolved.

Changing an Opinion

The opinion must carry the caveat, "and subject to revision if additional information is received in the future," Sometimes new information later comes to light, which necessitates changing an opinion. New information may arise in depositions or in an opposing expert's report. If the new information changes the engineer's original opinion, he must not hesitate to do so, perhaps in the form of a supplemental report.

Note

1 The decision-making principle attributed to William of Ockham, the 19th-century Scottish metaphysical philosopher, who may have been quoting a 14th-century English monk. The name is also sometimes spelled "Occam." "Razor" means a technique for finely slicing.

Afterword

It is probably apparent from reading this book that the successful practice of forensic mechanical engineering requires a wide diversity of technical knowledge. However, another important requirement is imagination. Not the ability to fantasize, but rather the ability to think outside the box, as the saying goes.

Some cases can seem particularly challenging – to seemingly defy explanation. It is up to the forensic mechanical engineer to find the explanation. Sometimes the key is asking the right question when other people have been asking the wrong one. For instance, in the case example about recurrent leaking joints in plastic water pipes, other people persisted in asking why the bolts were not tightened enough. In fact, the correct question was why the bolts repeatedly loosened despite having been properly retightened many times. This led to asking why any bolts loosen, which led to determining that the plastic surfaces were deforming at the high temperature of the water in the system.

Similar examples are described herein. Instead of asking why the shower leaked, the right question was when did the shower leak – only sometimes or every time it was used. Instead of asking why the cooking torch leaked, the right question was, what was the source of ignition if the torch showed no evidence of internal heat damage? Instead of asking what was wrong with the moaning boiler, the right question was why the boiler emitted the sound despite operating perfectly. In each case, asking the right question was the key to finding the correct answer.

The ability to ask the right question cannot be learned in school. Experience is the only teacher. A forensic engineer cannot give up when an explanation seems impossible at first. Gaining experience includes observing more experienced forensic engineers during joint investigations, doing research into similar phenomena, and yes, having an opinion shot down by opposing experts.

DOI: 10.1201/9781003684213-16

Index

For Product Safety Concerns and Information please contact our EU representative GPSR@taylorandfrancis.com Taylor & Francis Verlag GmbH, Kaufingerstraße 24, 80331 München, Germany

Printed and bound by CPI Group (UK) Ltd, Croydon, CR0 4YY
14/09/2026
02216223-0002